EFFECTIVE INTERPERSONAL AND TEAM COMMUNICATION SKILLS FOR ENGINEERS

IEEE Press
445 Hoes Lane
Piscataway, NJ 08854

EFFECTIVE INTERPERSONAL AND TEAM COMMUNICATION SKILLS FOR ENGINEERS

CLIFFORD A. WHITCOMB

LESLIE E. WHITCOMB

IEEE PRESS

A John Wiley & Sons, Inc., Publication

For general information on our other products and services or for technical support, please contact our Customer Care Department within the United States at (800) 762-2974, outside the United States at (317) 572-3993 or fax (317) 572-4002.

Wiley also publishes its books in a variety of electronic formats. Some content that appears in print may not be available in electronic formats. For more information about Wiley products, visit our web site at www.wiley.com.

Library of Congress Cataloging-in-Publication Data:

Whitcomb, Clifford A., 1954-
  Effective communication principles for engineers / Clifford A. Whitcomb, Leslie E. Whitcomb.
    pages cm
  ISBN 978-1-118-31709-9 (pbk.)
1. Communication of technical information. 2. Soft skills. 3. Engineering. I. Whitcomb, Leslie E., 1959- II. Title.
  T10.5.W483 2013
  620.001'4—dc23

                                                                                    2012027541

10 9 8 7 6 5 4 3 2 1

For Our Children

CONTENTS

SECTION II TAKING IT TO WORK 51

8 I, YOU, AND THE TEAM 53

PREFACE

This book is about learning effective interpersonal and team communication skills that are useful for engineers in the practice of their profession. Examples and exercises help you learn how to put together the basic units of effective engineering communication. Learning these basic units called microskills of communication, to use in your practice of engineering gives you options for handling issues that arise. Classic examples of these issues include moments when you are stuck with a project task that presents seemingly unresolvable technical issues or when you are stuck with a teammate who simply will not perform, or whose performance disrupts your own. In the process of learning how to handle these situations you will become an effective engineering communicator and you will be a better engineer. You will learn how to engage others. You will learn how to listen to others. You will learn how to manage conflict and influence others in highly constructive, repeatable communication exchanges.

The engineering field you have chosen as a profession holds as a primary purpose the benefit of society. The professional societies and tenets that will guide and bound your practice hold ethics, societal benefit, and the improvement of engineering effectiveness as their foundation. Our book guides you in the development of a significant new benefit to both your own profession and society. We give you this potential through our invitation to you to participate in a cutting edge engineering innovation—a skill set for effective interpersonal and team communication.

Throughout its history engineering has been nurtured by inventors and innovators who could see beyond the current limits of their field in order to create opportunities for social benefit. The steam engine was seen as a gadget that would never compete with the power of a horse. The computer was viewed as a sideline in relation to the real work that could be done by mechanical machines. Engineers saw beyond initial limitations and pulled together seemingly irrelevant and potentially disastrous elements to transform these nascent opportunities into full functioning contributions. They engineered these elements effectively and created technologies that contributed profoundly to benefit society for generation after generation.

Technical and nontechnical interpersonal communication is currently perceived as an almost irrelevant and minor component of the engineering process and engineering education—given short shrift even though it is continually required by professional societies and accrediting bodies, such as the Accreditation Board for Engineering and Technology (ABET). The ABET EAC 2010 Criterion 3 Student Outcomes lists several aspects for successful engineering education:

(a) An ability to apply knowledge of mathematics, science, and engineering.

(b) An ability to design and conduct experiments, as well as to analyze and interpret data.

(c) An ability to design a system, component, or process to meet desired needs within realistic constraints, such as economic, environmental, social, political, ethical, health and safety, manufacturability, and sustainability.

(d) An ability to function on multidisciplinary teams.

(e) An ability to identify, formulate, and solve engineering problems.

(f) An understanding of professional and ethical responsibility.

(g) An ability to communicate effectively.

(h) The broad education necessary to understand the impact of engineering solutions in a global, economic, societal, and environmental context.

(i) A recognition of the need for, and an ability to engage in, life-long learning.

(j) A knowledge of contemporary issues.

(k) An ability to use the techniques, skills, and modern engineering tools necessary for engineering practice.

Outcomes (d), (f), and (g) include aspects that require a learner's curriculum to address development of an understanding of professional responsibility, working on teams, and communications. These "soft skills", now sometimes referred to as professional skills, are given minimal space in already crowded engineering education curricula because they are often the hardest to teach, to learn, and to assess. The skills related to these elements are not necessarily best learned through classroom lecture, but through practice in authentic engineering contexts, such as capstone design projects. They are crucial none-the-less. Their foundational quality is also highlighted by the United Nations Educational, Scientific, and Cultural Organization (UNESCO) which has defined four pillars of education.

- Learning to know
- Learning to do
- Learning to live together
- Learning to be

The skills related to interpersonal communications primarily fit the intent of the Learning to be—which includes all aspects of human development— they also directly support Learning to live together.

Taking this global perspective and translating it into high quality engineering educational deliveries, the International Conceive, Design, Implement, and Operate (CDIO) initiative has sought to bring a more holistic view for the education and development of engineers.

There is a growing recognition that young engineers must possess a wide array of personal, interpersonal, and system building knowledge and skills that will allow them to function in real engineering teams and to produce real products and systems, meeting enterprise and societal needs.

(Crawley et al., 2011)

The CDIO initiative defines a syllabus for engineering education that addresses a broad span of competencies, technical and non-technical, for engineers, that should address

Specific, detailed learning outcomes for personal and interpersonal skills, and product, process, and system building skills, as well as disciplinary knowledge, consistent with program goals and validated by program stakeholders.

(Crawley et al., 2007)

In addition to being important for global competencies, communication is often cited as one of the most highly desired and important traits of a successful engineer in the US defense workforce. Figure 1 shows the results

Scale 1–5;5 = Very proficient, very mission critical

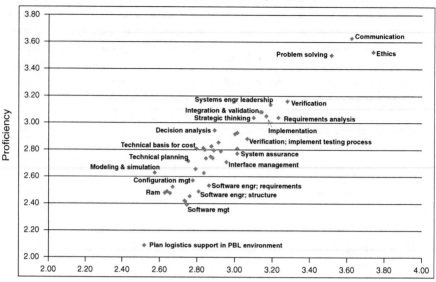

Mission criticality

FIGURE 1 Survey of the US Department of Defense (DoD) SPRDE workforce, with Mission Criticality versus the Proficiency Desired for the 29 identified Systems Engineering Competencies. These results show that next to Professional Ethics, Communication is the single competency desired that simultaneously requires the highest level of proficiency and is the most mission critical of any of the 29 engineering competencies surveyed.

of a 2010 survey of the US Department of Defense 38,000-member systems planning, research, development, and engineering (SPRDE) workforce (Center for Naval Analyses, 2011).

The results show that of 29 engineering related competencies included in the survey, the respondents desired a *Communication competency that requires the highest level of Proficiency and is one of the most Mission Critical skills for conducting engineering for SPRDE in DoD.* These results mean that next to professional Ethics, Communication is the single competency that simultaneously requires the highest level of proficiency and is the most mission critical of any of the 29 engineering competencies surveyed.

Another survey was conducted to determine levels of proficiency desired based on the CDIO syllabus. The results of the survey are shown in Figure 2 (Niewoehner, 2011).

The survey measured senior systems engineers' (those with hiring authority within their organization) responses to levels of desired proficiency for new hires and mid-career engineers with respect to categories of the CDIO syllabus. Systems engineers were asked, "At what levels of proficiency is it expected that a hired SE perform?" The levels of proficiency defined ranged

Desired skill proficiencies

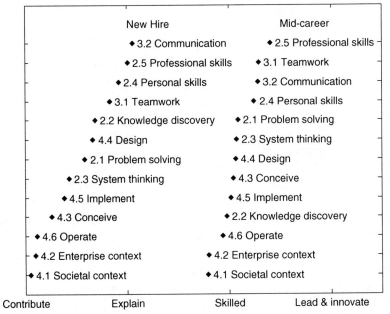

FIGURE 2 Results summary of the CDIO-based Survey of the Six Naval System Commands. The skills stack shows the importance of addressing the skills not necessarily related to any specific engineering discipline *per se*, for example, the ability to communicate, have professional skills (a sense of ethics, equity, and other responsibilities), personal skills (productive attitudes, the ability to think and learn), and to work on a team, are highly valued for new hire engineers, and continue to be some of the most important skills for mid-career engineers. (From Niewoeher, 2011)

from a low of "contribute" to the process all the way through the ability to "lead and innovate." The importance of addressing the skills not necessarily related to any specific engineering discipline *per se*, for example, the ability to communicate, have professional skills (a sense of ethics, equity, and other responsibilities), personal skills (productive attitudes, the ability to think and learn), and to work on a team, are highly valued for new hire engineers, and continue to be some of the most important skills for mid-career engineers. Notable as well is that by mid-career, engineers are expected to improve proficiency across the board in all areas.

What these two survey results do not show is that there are currently no dependable texts or development models available for interpersonal and technical verbal and nonverbal communication skills in engineering settings to benefit both entry level and experienced engineering professionals. There are no maps or charts—that show you as an engineer how to be a successful and proficient interpersonal communicator in both technical and non-technical settings.

We are offering you a model for that crucial competency. We are inviting you to be an innovator in your field. We are inviting you to take disparate components and opposing dynamics and explore new realms of possibility with these elements in hand. We are inviting you to add an exciting, fluid process—the multimodal properties and synergistic dynamics of human communication—into your technical engineering expertise to contribute to the field of engineering. We are inviting you to improve your capacity to benefit society.

Clifford Whitcomb, Ph.D., CSEP is a Professor and Chair of the Systems Engineering Department at the Naval Postgraduate School, Monterey, California. Dr. Whitcomb's research interests include model-based systems engineering for enterprise systems and systems-of-systems. He was previously the Northrop Grumman Ship Systems Endowed Chair in Shipbuilding and Engineering in the School of Naval Architecture and Marine Engineering at the University of New Orleans, a senior lecturer in the System Design and Management (SDM) program at MIT, and an Associate Professor in the Ocean Engineering Department, also at MIT. He earned his bachelors degree in engineering from the University of Washington, two masters degrees, one in Naval Engineering and one in Electrical Engineering and Computer Science, both from MIT, and a Ph.D. in Mechanical Engineering from the University of Maryland. He is a Certified Systems Engineering Professional (CSEP), and has served on the International Council on Systems Engineering (INCOSE) Board of Directors.

Dr. Whitcomb's career has spanned defense, industry, and academic settings. Throughout his career as a naval officer, submarine designer, researcher, professor, and academic chair in a diversity of settings, he has sought technical excellence and organizational innovation. This drive has led him to an understanding of the need for multimodal adaptive proficiencies in the education of engineers and the development of their professional competencies. He has found the development of the entire range of engineering proficiencies to be necessary because current engineering settings are impacted daily by innovations and pressures that involve social, technical, economic, political, and ecological shifts occurring across global networks of practice. He has found that keeping up with a fellow competitor takes technical excellence, and that keeping up with a multicultural, transnational, ever shifting enterprise system of competitors takes more than technical excellence. It takes technical excellence that is embedded in interpersonal competencies that support high-level adaptive responses to societal challenges.

Leslie Whitcomb, M.Sc. Applied Organic Psychology, Ph.D. Candidate, ABD, Sensory Integrative Ecopsychology, has been facilitating communication in a diversity of settings for twenty-five years. She has been a family educator, family counselor and communication facilitator, a teacher and curriculum developer at the graduate level, a curriculum developer and workshop leader in Applied Organic Psychology for Executive Management, Multicultural Education and Social Services professional trainings, and a research assistant in Clinical Psychology concentrations of non-verbal communication coding and sensory cue reciprocity in transgenerational family communications.

Through this breadth and depth of experience, Leslie has learned that communication is both shaped uniquely by the setting in which it occurs, and that communication has universal qualities that are possible to master across multimodal settings. Writing this book gives her an opportunity to expand the boundaries of both relational and engineering based communication capabilities. Leslie is delighted to participate in this endeavor and hopes that it enhances your professional and personal experience of communication as an engineer.

Feedback Solicitation
We would like to know what happens when you embark on this exploration. We invite your feedback to help us build a better educational and professional development process for engineers in the global engineering enterprise. Join us on Facebook or our group, Effective Interpersonal and Team Communications on LinkedIn or via e-mail at info@intentionalexchanges.com.

ACKNOWLEDGMENTS

Clifford would like to thank his colleagues in teaching and research. Their shared efforts demonstrated for him that effective interpersonal and technical engineering communication improves the practice of engineering. He would especially like to thank his colleagues involved with the project-based learning initiative in the Department of Systems Engineering at the Naval Postgraduate School, Dr. Diana Angelis, Gary Langford, Greg Miller, and Mark Stevens. Their commitment to being on the front lines of pedagogical and professional change—with all the challenge, risk, achievement, and reward that this entails—provided important inspiration to address the core learning in this book. They made a central contribution to this much-needed resource for Dr. Whitcomb's peers and students. He would also like to thank Ali Rogers, NPS Director of Faculty Development, and Dr. Ed Crawley, from MIT, who provided much inspiration for his taking a critical look at engineering education, and for providing motivation to constantly seek to improve it.

Leslie would like to thank Dr. Andrea Di Benedetto and Dr. Susan Theberge, her colleagues in mastering the intricacies of communication in educational and counseling settings. Leslie would also like to thank Dr. Susan Eliot, Dr. Nancy Baker, and Dr. Michael Cohen, her mentors in psychology and ecopsychology. These practitioners have evolved work in pioneering realms of indigenous, multicultural and ecopsychological research and practice. In doing so, they have modeled for Ms. Whitcomb that communication can be both context specific and transfluential.

Together, we would like to acknowledge the kind and generous professionals who took time and focus away from their own work to reflect upon the manuscript in development. We thank Michael Archer for his work in creating our graphic designs. Mary Vizzini, Barbara Berlitz and Ann Shows offered helpful expertise from communications, critical thinking and technical writing perspectives. Dr. Ben Roberts gave positive feedback and helpful reflection based on his insights as an Organizational Psychologist. We would especially like to thank Dr. Rob Niewoehner and his son, RJ. They are both engineers and they are a father and son who have communicated successfully across generational boundaries. Their feedback was both absolutely on target from an editorial perspective and it was a model of successful engineering interpersonal and technical communication in action.

We would like to thank everyone at Wiley who guided us through this venture. We would like to thank Tai Soda for being receptive to our original idea. Most especially, we would like to thank Mary Hatcher for expert, on-point editorial guidance through out the process. Our book would not be what it is without the contribution of the complete Wiley team.

Finally, we would like to acknowledge the students and colleagues who we hope will benefit from this book. Working with you has inspired us to create and offer this resource. We hope it benefits you as members of society and as engineers.

SECTION I

LEARNING THE BASICS

CHAPTER 1

 LEARNING TO DRIVE YOUR COMMUNICATIONS

Learning to drive your communications means you control powerful forces that impact interpersonal and technical clarity in getting your point across in engineering settings.

Learning to Drive Your Communications

Remember when you learned to drive a car?

You had used your arms and legs before you began driving. You had learned to coordinate your vision with your sensory motor choices before you ever got behind the wheel. But when learning to drive, you were using these capabilities in a context that asked you to grow a new integration of your skills.

If you remember the sharp stopping of the car when you first learned to apply pressure to brakes . . .

If you remember the hard scrape of the wheel grinding the curb when you learned to parallel park . . .

Effective Interpersonal and Team Communication Skills for Engineers, by Clifford A. Whitcomb and Leslie E. Whitcomb.
© 2013 by The Institute of Electrical and Electronics Engineers, Inc. Published by 2013 John Wiley & Sons, Inc.

. . . then you have a sense of what it will feel like to practice the engineering communication skills you will learn in this book.

Prepare to feel disoriented.

It means you are actually learning to drive the intention of your communications rather than being driven by them. You are learning to understand your own thoughts, feelings, and behaviors as a communicator. You can master these dynamics, just like you eventually learned to drive a car like it was second nature, through learning our *Communication Microskills Model* Figure 1.1.

COMMUNICATION MICROSKILLS MODEL

Microskills Definition

Microskills are elemental, or subunits, of communication skills. These are labeled on tabs, analogous to DNA base pairs, in our model Figure 1.1. Learning to use them individually and in a variety of combinations allows you to build your DNA of holistic skills as a communicator in a diversity of engineering tasks and settings.

You develop these microskill tabs (sub-units) one-by-one, then combine them into effective, fluent communication skills. You use these distinct subunits, the microskill tabs, alone and in a variety of combinations— creating shared communication strands of information exchange with others, to listen, to anticipate, to predict, to respond, and to become a better engineer.

WHY ARE MICROSKILLS IMPORTANT AS A BASIS FOR COMMUNICATION IN ENGINEERING?

Learning basic units of communication for practice in your profession is important because engineering communication is a complex mix of social dynamics (think about the diversity of people in an organizational context and how they must all work together to develop services and products) and technical expertise (think about analyzing and designing the physical characteristics of products).

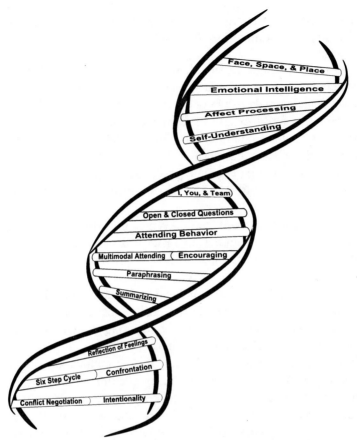

FIGURE 1.1 Communication Microskills Model. Think of the various microskill subunits as analogous to base pairs in a strand of DNA. The microskill base pairs are the essential building blocks for the development of your holistic DNA of information exchange capability in interpersonal and technical engineering settings. We label these microskills as tabs in the model given to you in this book.

The microskill tabs on our model simplify interpersonal and technical exchanges that occur in complex engineering situations. Successful use of microskill tabs individually and in a variety of combinations ensures that engineering ideas, designs, and operations are accurately expressed and received by all professionals involved.

This usage is important to learn because the alternate is letting communication remain an unskilled aspect of engineering practice—and then dealing with the consequences of negative project outcomes that correlate with deficits in professional engineering communication proficiencies.

Negative Project Outcomes Correlate with Deficits in Professional Engineering Communication Proficiencies

Impacts of deficits in professional engineering communication proficiencies were starkly demonstrated in the space shuttle *Challenger* disaster. Top-level decision makers had not been accurately informed of problems with O-ring seals and external sheath joins on the shuttle, even as the countdown to launch commenced. Concerns about the impact of cold weather on these elements during servicing delays on the morning of launch were not communicated adequately by engineers on the project, nor were they given full attention by NASA officials immediately before launch.

Communication that could have prevented tragedy was not expressed or received with enough accuracy to make a difference. The shuttle was engulfed in flames soon after launch, killing all on board, due to an O-ring failure and sheath leaks. Investigations after the event supported the conclusion that flawed interpersonal and technical engineering communication was a significant factor in the deaths of seven astronauts and in subsequent discontinuation of major engineering design contributions to space exploration (McDonald and Hansen, 2009).

While this consequence of deficits in communication proficiencies is especially dramatic, it occurred on a spectrum of interpersonal and technical communication miscues that are a prevalent aspect of the engineering profession. Design failures and subsequent social and professional consequences happen regularly in development and post-completion stages of engineering projects. Failures in interpersonal and technical communication in engineering settings play a significant role in these issues.

Learning to use the Communication Microskills Model is an excellent way to help prevent these issues from impacting your professional practice as an engineer.

HOW DO MICROSKILLS WORK?

When you look at our Communication Microskills Model, you can see the microskill tabs occurring individually and in connection with all the other skills in the helix.

The microskills are similar to the encoding on a strand of DNA because each microskill carries information that shapes its own function and relates to overall communication functions that become holistic interactions.

For example, the microskill tab of I, You, and Team statements helps you make clear statements that are not over-personalized nor overly technical and generalized. Even better, when this microskill tab is combined with the microskill tab of attentive listening behaviors (Attending Behaviors) you are expressing *and* hearing important interpersonal and technical content concerning engineering tasks and project developments.

HOW WILL I LEARN THE MICROSKILLS?

Each labeled microskill tab will be explained in detail in this book. You will be given examples and questions with each microskill tab to demonstrate how to generalize this knowledge and practice these skills in real engineering settings.

Each microskill tab will also be demonstrated in combination with others, allowing you to form new and ever-expanding communication proficiency structures that now have their own properties that fit a variety of engineering contexts. Scenarios and dialogs in the chapters show you how individual subunits combine and work together in holistic interpersonal and technical engineering information exchanges.

These exchanges are invaluable in moving your projects forward and preventing them from being mired in emotions and behaviors that block technical progress.

WHAT'S IN IT FOR ME?

The examples and exercises help you learn how to put together the pieces of effective engineering communication. You will become an effective engineering communicator and you will be a better engineer. You will learn how to engage others. You will learn how to listen to others. You will learn how to manage conflict and influence others in highly constructive, repeatable communication exchanges.

Throughout this book, reading dialog demonstrates the presence of these skills and prepares you to recognize their presence in yourself and others. Defining them and then understanding their contextual integration provides deeper learning to understand them more completely.

WHY THIS WORKS

The basis of why this will work for you is because our model represents elemental communication basics that form the foundation of powerfully effective engineering communication proficiencies.

Microskills of communication become well-defined engineering skills when you learn to use them to

- get the *attention* of others,
- *engage* others with you,
- get your *point across*,
- *persuade* others,

to move engineering functions toward engineering outcomes that are accurate and successful.

Use these microskills for

- *providing the controls* for the engineering communications that *you are driving*
- *allowing yourself to think, feel, and respond* with interpersonal *intentionality* and technical excellence.

THE IMPORTANCE OF A PRACTICE-BASED MODEL

The engineering communication microskills in this book will serve you well if you let them, and if you take the time to use them. *You unlock and increase their potential each time you practice them in authentic situations.* Upon completion of this book, you will be able to

- learn and master intentional engineering communication through microskills,
- understand communication microskills,
- apply communication microskills in engineering contexts, both with individuals and teams,
- draw out individual and team issues and problems through the use of a basic listening sequence,
- develop strategies leading to individual and team change and action,
- learn and master the influencing microskills of confrontation and interpretation,
- listen, influence, and structure effective communication exchanges,
- predict the likely impact of your efforts in structuring communication exchanges.

A recommended way to use this book is to first focus on a single microskill. Read about it and develop a cognitive understanding and then, *practice*. You can only expect to become better at communication if you actively practice. This is more than simply reading about the microskills and doing them once, and thinking that this is easy. You need to continually practice in authentic situations to understand the effectiveness of your application of the microskills. The examples and exercises teach you how to put together the pieces of effective engineering communication to make this learning through practice happen.

The initial chapters in this book cover *Learning the Basics* related to all communications, from Shared Communications to Your Natural Style, Self-Understanding, Emotional Intelligence, and Affect Processing. All of these integrate to give you a crucial understanding of the interpersonal field in which all engineering technical and nontechnical exchanges occur.

Chapters in the section, *Taking it to Work*, help you to build the basics, to form a bridge between you as a *technical engineering* communicator and you as an *interpersonal* communicator.

Practicing these skills repeatedly points you toward intentionality in your technical and nontechnical communications. This intentionality is expanded in the third learning level of the book, putting you on a path toward intentional competency.

Chapters in the section, *Making it Real*, model intentional competency for you. They show you how to keep technical and interpersonal communication exchanges constructive even when colleagues have widely divergent opinions on technical approaches or interpersonal styles and even when projects run into technical or schedule glitches.

Chapters in the section, *Taking the Lead*, provide you with additional skills to practice when worst-case scenarios occur in engineering settings. Knowing how to deal with these scenarios gives you an *engineering communication intentionality* that makes you not only a valued player but a leader in your profession. Some of these scenarios involve confrontation and include situations like a peer or teammate who won't cooperate or perform, a supervisor who has no management skills, or a project design that evolves intractable technical and schedule glitches near the end of a project cycle.

The chapters on using communication skills throughout *Conflict Negotiation* lead you toward this mastery. They give powerful fluency and effectiveness to your engineering communications. These chapters model engineers communicating effectively producing effective engineering solutions—even during moments of conflict, intense creativity, interpersonal differences, and intractable project-related troubleshooting.

When you have practiced these skills in authentic engineering settings and experienced their positive impact on your communications and engineering outcomes, you are an *intentional engineering communicator*. You can now intentionally drive engineering tasks and goals and fulfill engineering proficiency potentials in ways that meet and exceed expectations consistently.

The final level in mastering effective engineering communication skills to produce effective engineering outcomes is *intentionally modeling* engineering communication competency, not just for yourself, but for those around you—and you become an Intentional Engineer. Because you are now *driving* your engineering communications rather than being *driven* by them, you are driving your responses to the technical and interpersonal communication inputs of others rather than being driven by them. Your technical and interpersonal content, your emotional and behavioral responses to others in your communication content, becomes a model for, and teaches others, how to drive effective communication and effective engineering outcomes, as well.

CHAPTER 2

WHAT DOES IT MEAN TO BE AN EFFECTIVE ENGINEERING COMMUNICATOR?

The Space, Face, and Place Spectrum is the microskill tab in our Communication Microskills Model that helps you understand shared contexts for interpersonal and technical engineering communication exchanges.

The following dialog between two engineers will introduce you to this initial, self-placement skill of engineering communication—identifying where your technical and interpersonal exchange is happening and adjusting your content to be congruent with the shared setting in which you are communicating and relevant to the people with whom you are communicating. This is the Space, Face, and Place Spectrum for interpersonal and technical content in engineering communications.

Shared Basis of Communication Exchange Scenario

Dex and Michelle are creating an app for a smartphone. They are trying to merge pictures, tweets, campsite maps, and activity suggestions from Hiking Club outings taken with friends to add to their club's outing resources. They are stuck on a particular section of code and despite double-checking, several times even, and reviewing the logic repeatedly— the app is not functioning. Dex bumps into Michelle in the quad outside the lab and he is glad to have a chance to check in with her about progress as

Effective Interpersonal and Team Communication Skills for Engineers, by Clifford A. Whitcomb and Leslie E. Whitcomb.
© 2013 by The Institute of Electrical and Electronics Engineers, Inc. Published by 2013 John Wiley & Sons, Inc.

time pressure is mounting. Michelle is on her way to another meeting and the following dialog ensues.

Dex, "Oh hey, Michelle! Did you make any progress on the programming glitch last night?"

(Dex didn't fully recognize the *Space* between himself and Michelle. He did not acknowledge that bumping into Michelle in a neutral setting outside of a lab space or office space might need to include some social context. Instead, he went right to purely technical content. Michelle could miss important technical content because she may not be in full on technical mode in this particular setting.)

Michelle, "Well, not after we finished late yesterday afternoon, anyway. My Nana is in town from Singapore and my parents and my brother are here. We all went out for a family dinner."

(Michelle is giving Dex a cue as to how to balance their *Face* roles at this time, they are having both a social and a technical moment because they are at work but they are not in an actual work place setting—so a little more purely social interaction might be appropriate. Michelle is trying to let Dex know she is not ready to accurately receive technical content at this point.)

Dex, "Nice. Hope she's feeling better than her last trip over. But, I wanted to tell you I think I had a breakthrough in the coding. I found that if I did the data query too soon in the sequence it wouldn't be received . . . "

(Dexter is giving a technically pertinent explanation of his code breakthrough but failing to notice Michelle's shifting posture, eyes rapidly glancing at the door to her office building and increasingly glazed look as she tries to take in details that are best integrated with a computer screen in front of the viewer. He is not in context of their shared *Place*. He is unintentionally forcing Michelle to attempt to have a technical discussion without the tools to adequately do so. Any technical feedback she could give him is being lost in the fuzzy Space, Face, and Place boundaries going on in this engineering communication exchange.)

Michelle—holding up her hand and smiling, "Dex, I know you must be super psyched to figure this thing out but I've got to be at a meeting that started 10 min ago. How about you save this 'til we meet in the lab at 3?"

Dex, shrugs and smiles with embarrassment, "Glug. Sorry Michelle. I'll meet you in the lab at 3. And I do want to hear if you took your family to that great restaurant you talked about last week. See you then."

Michelle continued to remember where she was on the *Space, Face, and Place Spectrum.* Dex followed her cues and brought them back to a balanced interaction that matched technical need, professional and interpersonal roles, and their setting.

SHARED BASIS OF ENGINEERING COMMUNICATION EXCHANGES DEFINED

This dialog highlights the importance of understanding the basis of shared engineering communication exchanges. This basis consists of

- the physical space within and directly around speakers as individuals,
- the physical space within and directly around the speakers as a communicating system,
- patterns of hearing, seeing, speaking, and posturing exchanged between the speakers,
- contextual relevance of an exchange—a balance of technical and interpersonal communication skills and a match between content and setting,
- the shared space present in all technical and nontechnical communication exchanges that happen in the field of engineering.

Every successful communication you will engage in as an engineer will begin with the basics of understanding and managing this shared physical space and these shared patterns of speaking and listening in social settings and contexts of engineering. This shared space is the *Technical and Interpersonal Space, Face, and Place Spectrum.*

Technical and Interpersonal Space, Face, and Place Spectrum

The spectrum components are:

Range One—Primarily Social

Team formation, job interviews, project start-ups, client interviews, conference meet and greets, professional networking that happen in classrooms, labs, meeting rooms, and conference areas.

On this part of the spectrum

- Your conversation choices are general—weather, common interests, sports scores.
- Your goal is to keep yourself comfortable and make others comfortable.

Microskills that aid in making this engineering communication successful are

- attending behaviors, open and closed questions, and multimodal attending.

You will learn these microskills in subsequent chapters of the book.

Range Two—A Balance of Social and Technical

Team meetings, project report sessions, client presentations, or conference presentations that happen in labs, classrooms, coffee rooms, offices, conference rooms, and halls.
On this part of the spectrum

- Your conversations and presentations are about a specific engineering topic, goal, design, or operation.
- Your delivery holds both technical and interpersonal content because you are communicating concepts that must attend to both technical aspects and interpersonal dynamics.

The shared space holds clues about patterns of hearing, seeing, speaking, and posturing of yourself and those around you that would indicate signs of confusion, misunderstanding, emotional reactions, or technical disagreements.

Range Three—Purely Technical

Class lectures, team design meetings, individual project design meetings, presentations to teammates, lab workouts, data sharing, and design analyses that happen purely in professional workspaces.
At this point on the spectrum you are

- focusing on conveying accurate technical detail to peers, supervisors, and clients,
- ensuring your details are accurately understood,
- listening carefully to take in the technical details others are conveying to you,
- receiving detailed feedback in a way that allows the speaker to fully explain their position.

SPACE, FACE, AND PLACE SPECTRUM DEFINED

Space refers to the shared social and/or professional context that exists between the people who are communicating.

Face refers to who you are talking to—a peer, a supervisor, a technical group, for example.

Place refers to where the communication is happening, in the team meeting room, a conference venue, video teleconference, a boardroom, or a coffee shop, for example. Social cushioning happens on this spectrum—moving from being the primary focus of an exchange through being balanced in the exchange into being almost completely in the background of an exchange. The primary focus of the exchange can shift along this spectrum.

CONTEXTUAL INTEGRATION OF SHARED COMMUNICATION SPACES

Engineering communication occurs in a context of shared, primary, life resources. This shared context grows from the fact that you and your colleagues are earning your living together and you are creating products that serve society. This means that although your interactions will be primarily technical, they will consistently impact *social*, primary resources (salary, promotion, and long-term career stability). Perhaps even more important, although your interactions in a shared social context will be primarily technical, they will impact *public, social* resources (safe bridges, functional industrial, or tech equipment) for people who use engineered products.

Understanding this context becomes very important because

Shared social resources generate shared social dynamics so communications happening in shared social contexts of engineering will involve technical concepts *and* interpersonal dynamics.

Assessing and using microskills of communication enables you to balance this reality in your engineering communications. The first step in maintaining that balance is identifying your existing natural style of communication.

CHECK IN

Use this rubric to check in on how well you are developing the Face, Space, and Place microskills.

	1 = Not Attained	3 = Satisfactory	5 = Outstanding
Space, Face, and Place Spectrum	Content and/or delivery does not match social and/or professional context, who is being addressed, or place where the communication is happening	Content and delivery adequately matches social and professional context, who is being addressed, and place where the communication is happening. Maintaining flows of constructive communication and influencing outcomes is not yet present	Content and delivery match social and professional context, who is being addressed, and place where the communication is happening. Constructive flows of communication are maintained and outcomes are effectively influenced.

CHAPTER 3

YOUR NATURAL STYLE OF COMMUNICATION

Your natural style of communication is the microskill tab in our Communication Microskills Model that helps you identify your own unique strengths as an interpersonal and technical engineering communicator.

To help you think about your natural style of communication, here is an exchange that occurred between two engineers about a project. These two engineers have two very different communication styles, yet their communication was still highly successful on both technical and interpersonal levels.

WHAT ARE THE STRENGTHS OF YOUR NATURAL STYLE?

An engineer from a Silicon Valley technology firm is breakfasting with a colleague from Asia. They are meeting in San Jose, California, to address the final stages of development for an end-to-end systems approach offered by their company. It will deliver desktop virtualization and virtual workspaces that will provide a superior integrated voice and video user experience.

Their teams have had to work together to design systems and modules that reduce risk and simplify deployments through a fully tested system that combines data center, network, and collaboration architectures together.

Effective Interpersonal and Team Communication Skills for Engineers, by Clifford A. Whitcomb and Leslie E. Whitcomb.
© 2013 by The Institute of Electrical and Electronics Engineers, Inc. Published by 2013 John Wiley & Sons, Inc.

They are coordinating final presentations for product exhibition at a major tech show in Hong Kong.

Don, the lead engineer, is warm and expressive and talks at a rapid pace, using lots of eye contact, smiles, and humor as he gets his point across. He is senior in the organization but Lin, the systems engineer, carries a lot of weight in the project so his opinion is valued. Lin has a much more subtle and quiet natural communication style. You'd want him next to you in any critical meeting. He doesn't miss a thing.

During their interchange, Don puts out a lot of wattage while he speaks but he doesn't bulldoze over Lin. He leaves long enough pauses so Lin could easily interrupt or disagree. And when Lin does speak, Don gets still and quiet. Continuing to smile and be open, but communicating solely with his body posture, not his words.

Lin is a quiet listener. He shows he is listening by keeping his head down, turned slightly to the side and very occasionally nodding his head.

How does Don know Lin is listening even though his head is down? Because Lin's shoulders and torso are fully facing Don. Because when Lin asks a question it is precise and gives back specific, insightful details of Don's content that only someone listening very carefully could have received.

Why doesn't Don feel insulted that Lin is not looking at him? Smiling back? Gesturing back? Because when Lin does respond, he responds with positive encouragers. Questions like, "So what should we do about this?", asked with a polite, inquiring, and encouraging tone. These go a long way toward meeting the intensity of Don's style with an intensity and clarity that is all Lin's own. The meeting concludes. Now both men smile and shake hands. They have had a successful communication, even though both their styles are so different.

Their communication was successful because they both used their natural styles to ensure content was delivered and clarity was achieved. Technical *and* interpersonal organizational content was delivered, understood, and integrated behaviorally, with follow on, by those involved in the communication. No one involved walked away feeling unclear, misunderstood, or with interpersonal weather (confusion, frustration, future concerns, feeling shut out, or over ruled) that will impact future communications.

NATURAL STYLE OF COMMUNICATION DEFINED

Your natural style of communication grew with you as you learned to walk, talk, play, and interact with the world. Your natural style is so basic to who you are that for most of the time it is not necessary to assess and examine how it works when you communicate. But when you want to make sure that your technical content reaches others in a way that they understand clearly enough to move engineering tasks forward, and when you want to make sure that your interpersonal dynamics do not create obstacles to engineering processes, it is helpful to recognize the basics of your natural style of communication. These include

Your posture: the way you hold your body in relationship to others as you speak—how open or closed you are in facing another, for example.

Your facial expression: the way your face moves and changes and shapes itself as you express words and as you listen to others.

Your voice tone: the way your voice shifts in volume, emotional emphasis, and rate of speech as you speak and listen.

Your word choice: do you use just a few words to convey complex content, or are you someone who needs to bring in a breadth of description and explanation in order to convey your meaning? Or do you fall somewhere in between on that spectrum?

There is no right or wrong in your natural style. Like a national park, it is a place you want to know and respect but not control and cut back. The point of identifying your style is to be able to smoothly track its impact on others so that your communications have your desired intent, rather than creating unintended static in your technical and interpersonal communications.

Try This

Think about your own natural style of communication in the following contexts.

Think of a situation where you have had to work in a technical or engineering situation and you engaged in a technical or engineering dialogue with someone, or with a group of teammates. Think about how you would describe a brief sketch of that scenario.

What were your thoughts? Were you able to express your thoughts in a way that others clearly understood and that met your expectations for the message you were trying to deliver? Think about one example here of something you said that was clearly stated and accurately heard by a peer or your teammates.

What were your feelings? Did you feel accurately heard and positively received? Did you feel resisted or encouraged? What parts of your body conveyed your own feelings to yourself? Ideas in your head? A tightening in your gut? A relaxation in your shoulders?

What existing strengths do you bring to developing your communication skills? Are you a good listener? Do you have a persuasive style that others respond to enthusiastically? Are you able to convey complex information in a way that is not intimidating to the layperson?

CONTEXTUAL INTEGRATION OF YOUR NATURAL STYLE

You were a communicator long before you were an engineer. Now you are both a communicator and an engineer. Everyone has a natural style of communication, developed over a lifetime. You do as well. So, to get the most out of this book, you can start by looking at yourself and respecting your own current competencies of communication. Equally important, you can look at yourself and assess where in your communication patterns you may need to develop new skills and strategies to improve on your own natural style. This is important because your natural style is the foundation from which you will communicate through out your career. But it may not work with everyone, or in every situation. You may need to shift, adapt, and change to become fully effective in the variety of situations that will confront you through out your professional practice.

Everyone has communication strengths and communication challenges. Knowing your strengths gives you a stable foundation from which to address your challenges. One of your greatest strengths can be the ability to take in multiple perspectives through feedback from others. This feedback about your own strengths and challenges from others ideally hones your engineering communication skills.

In becoming a more effective communicator, you will become a more effective engineer. Here's an example of an engineering approach that embodies both the need for effective communication with customers to engineer a product that meets their needs and the engineering results that create a good product design.

BuildSmart Program

Note: In this example, listening to the customer need (in **bold**) leads directly to engineering solutions (in *italics*).

Our BuildSmart Program has also **evolved by direct input from our owners—what works best and why, what could be improved on and why**. Truelux is always striving for excellence. We are immensely grateful to all our owners, many of whom are in touch on a regular basis contributing ideas, testimonials, and most importantly their recommendation of the Truelux Brand.

All Truelux 5th Wheelers are BuildSmart and **start with an in-depth consultation with our clients to determine exactly how they wish to use their 5th Wheeler and what tow vehicle they intend to use.** This *ultimately determines* the floorplan and *features* to be included. Fundamental in RV design excellence is functionality—allowing maximum use of available space, easy access to the fridge, cupboards and storage with ample work bench space either side of the sink and stove.

Note: What comes next is a way this highly successful company has taken a facility with interpersonal and technical communication to new levels of design that reflect a wonderful natural style of responsiveness to customer needs backed up by engineering excellence.

Truelux 5th Wheelers offer Ease of Hitching/Unhitching

Hitching is much easier with the tow ball clearly visible on the tray or in the tub (not under the car behind the bumper bar as with a caravan). It's generally at waist height and easily accessed to push home the locking pin. Some 5th Wheeler owners are retired women who independently hitch and unhitch their rig with ease, Rosanna below, thought up this nifty idea to keep the ball grease off her dog, Layla, by cutting the bottom out of a 1 lt Coke bottle she thought this was better than wasting a stubby holder!

The communication exchanges that lead to successful engineering outcomes include an ability to listen to what customers need, and communicate that to a team in order to meet the right market need with appropriately engineered products. This listening and interpersonal/technical balance in communication to find shared solutions happens easily when you achieve mastery in the use of engineering communication microskills.

CHAPTER 4

HOW SELF-UNDERSTANDING LEADS TO DEVELOPMENT OF EMOTIONAL INTELLIGENCE

Self-understanding is the microskill tab in our Communication Microskills Model that supports proficient usage of all subsequent skills.

> Self-reflection and its reward of self-awareness cannot be thought of as passive exercises, new era meditation, or soft science. They're absolutely essential.
> —Tjan (2012)

Project Communication Sequence—Self Understanding Leads to Emotional Intelligence, and Emotional Intelligence Leads to Improved Engineering Communication and Improved Outcomes

Note: Foundational microskills of self-understanding and emotional intelligence are in **bold**, individual behaviors that express proficiency in these skills are in *italics*.

Self-awareness—Dylan is in a Capstone Design Project selection session. His classmates are choosing their capstone projects for the year. He is aware that his shoulders are hunched, his breath is shallow, and he's staying quiet and hanging back from the group

---

Effective Interpersonal and Team Communication Skills for Engineers, by Clifford A. Whitcomb and Leslie E. Whitcomb.
© 2013 by The Institute of Electrical and Electronics Engineers, Inc. Published by 2013 John Wiley & Sons, Inc.

joshing and joking going on. (*Tracking physical changes that mark communication exchanges.*)

Self-regulation—Capstone Design Projects are a really important part of engineering education and are talked about often in Job Fair interviews at his school. The pressure is intense to show technical wizardry that is ahead of the curve in any project. Dylan knows he is a quiet, behind-the-scenes analysis guy. He's got a nail-biting level of angst at this meeting, hoping peer pressure won't make him choose a project in which he's not comfortable. He knows he needs to be a part of something that really suits his skills, so he's trying to keep out of the competitive, posturing banter and teasing the other students are doing as they trade ideas. (*Careful observation of technical content, nontechnical content and interpersonal feelings.*)

Self-motivation—Dylan really enjoys systems level analysis and is motivated to choose a project he doesn't mind spending hours and hours going over to get just right. He's motivated not to repeat a first year project experience that got a lot of attention from supervisors and program sponsors, but that put him constantly in the limelight to explain the technical details with respect to the design innovation. He feels like he lost about 20 pounds in blood, sweat, and tears getting through that and swore he'd show proficiency in his own comfort zone next time. (*Driving, instead of being driven by your response to technical and nontechnical interpersonal communication exchanges.*)

Empathy—Dylan watched teammates suffer through his stilted presentations and saw the discomfort on his advisors' faces when they had to give him fewer points than they knew he had worked for. He didn't want to go through that again.

Social attention and focus—Dylan walks over to the one advisor in the room who has a specialty in the kind of analysis where his own strength could be best used. He gets up his courage while he waits for a break in the advisor's attention, then starts to describe his project. The advisor nods her head, looks straight at him, closes her computer, and sits back in her chair. He sees some spark of interest coming from her in his direction. He really warms up and gives her a detailed idea for the project he'd like to do. He then clears his throat and explains he learned from the earlier project that out-front team presentation and leadership isn't his strength. He says he feels that this is what he is better at, and the accurate and excellent job he'll be able to do will reflect well on his efforts as a whole when capstone results are totaled for the team.

His advisor smiles and says he should go by what he learned the first time around. She can help him with resources and there is a lack of his type of technical skills for some of the project choices, so it should work out in the end. Dylan smiles and nods his head. His advisor's response makes the embarrassment of speaking up for himself more than worth it. Maybe he won't be able to take a lead role on the most cutting edge project with all that unwanted attention, but he'll do a great job and get the attention of an advisor who is strong in his eventual engineering technical focus. *(Managing the shared field of communication exchange so that technical and nontechnical interpersonal exchanges remain clear.)*

SELF-UNDERSTANDING MICROSKILLS DEFINED

As we learned from observing Dylan, self-understanding is built on self-awareness, self-regulation, self-motivation, empathy, and social attention and focus. When these microskill units are practiced through behaviors of

- self-tracking,
- careful observation of technical/interpersonal content,
- driving—rather than being driven by—communication exchange responses.

They become behaviors that express the following, defined microskills of self-understanding.

Self-Awareness

Self-awareness means you understand your strengths and limitations, your self-image, and how you feel about yourself. You can track your own communication responses and those of others accurately.

Self-Regulation

Self-regulation means handling your feelings, and not letting a situation get the better of you. When you have self-regulation under control, you can deal with challenging situations better.

Self-Motivation

Self-motivation means you have an inner drive to accomplish what is important to you. To achieve self-motivation, you must be persistent in achieving your goals, and not give up easily.

Empathy

Empathy means you are interested in others, and that you care about them. Empathetic understanding allows you to include others and see their perspective.

Social Attention and Focus

Social attention and focus allow you to use your self-understanding in balance with the communication styles of others. When used together, the above microskills can be considered to be a behavior of emotional intelligence.

All the microskills in this book translate self-understanding and emotionally intelligent behavior into effective engineering communication and effective technical engineering outcomes. This translation happens through the following skills:

- attending to your position on the space, face, and place spectrum in any given communication,
- tracking breath rate, muscle tone, voice tone, posture and facial expression changes in self and others,
- observing, carefully, the technical content, nontechnical content and the interpersonal weather within which this content is being transmitted and received,
- managing this shared field so the transmission and reception of technical and nontechnical communication stays clear,
- driving, instead of being driven by, your response to the technical content and shared interpersonal fields of engineering settings.

CONTEXTUAL INTEGRATION OF SELF-UNDERSTANDING

You now may probably be wondering what you have gotten into. You may be thinking, "You want me to understand my thoughts and feelings? I'd rather have a root canal without anesthesia." You may be thinking, "How do the professional skills of self-understanding and emotional intelligence impact my proficiency as an engineer?"

Self-understanding and emotional intelligence are crucial precursors to communication microskills usage in engineering settings because shared engineering tasks generate shared social dynamics—both interpersonal and technical. Simultaneously and equally important—social dynamics of engineering generate shared social emotions as well as shared engineering activities.

Shared social and emotional dynamics may not be primary in engineering professional settings but they are present in the communication feedback loops of your teams, companies, and enterprises and will impact design and budgetary choices. Emotions may not take the lead in engineering professional settings, but emotions are still present in your brain processes while you accomplish engineering activities.

This is a fundamental dynamic of communication for engineers because as an engineer

- You are a proactively contributing member of that social system and its unique communication needs, using both technical and interpersonal modes of communication effectively for the needs of that system.
- You constructively shape the flow of technical and interpersonal communications during your teamwork and project operations.
- You are able to find communication equilibrium in interpersonal and technical modes even though you are more likely *to not be* understood by your teammates, colleagues, and clients than you are *to be* understood by your teammates, colleagues, and clients.

These skills are amplified and operationalized in the microskill tabs in the Communication Microskills Model. We will take you through them step-by-step.

First, we will give you an example of emotional intelligence in action to help you begin integrating a crucial foundation of using all microskills— balancing your reason and your emotion when engineering challenges or professional setting challenges impact your interpersonal and technical communication content.

CHAPTER 5

DEVELOPING EMOTIONAL INTELLIGENCE

Emotional intelligence is the microskill tab in our Communication Micro-skills Model that gives you the basis upon which you learn to balance reason and emotion.

Emotional Intelligence in an Engineering Context Scenario

Jaida is in a team meeting to discuss final alterations for a model rocket launch in the Mojave Desert. This is a capstone project. Successful teams in this project often have members chosen to qualify for teaching assistant positions and tuition assistance.

Jaida's team chose a design and adapted it to meet specs given by the instructor. Jaida is not a socially confident speaker and has stayed very quiet in team meetings. She has fulfilled all her team tasks thoroughly and with more than above average excellence so she has the respect of her teammates. She is very worried because in initial trials the rocket motors did not perform as predicted and in some tests they failed altogether. The calculations and procedures have been checked and rechecked and team members have debugged many details, but still, no-go.

Effective Interpersonal and Team Communication Skills for Engineers, by Clifford A. Whitcomb and Leslie E. Whitcomb.
© 2013 by The Institute of Electrical and Electronics Engineers, Inc. Published by 2013 John Wiley & Sons, Inc.

Jaida feels intuitively that there is some mismatch between her team's rocket motor prediction model and the test experiment.

Speaking up about this issue means going head to head with Alex, the most vocal team member, and the one responsible for selecting that specific prediction model. She is torn between her need to help the team and her desire to avoid a public and potentially contentious exchange with Alex. Jaida chooses the following *emotionally intelligent behaviors* during the team meeting.

1. She identifies her emotions as she listens to Alex try to sell his selected prediction model to the perplexed and frustrated teammates around her—her emotions include, frustration, concern, getting very quiet and feeling her heart racing, and actually feeling some anger toward Alex as he keeps talking and talking and not letting team mates interject.

2. She notes that her shoulders are hunching and her breath is getting tight as she shrinks in her chair, and that she is not visually connecting with any of her teammates. She has stopped looking at others.

3. She observes that Alex is taking up more space in his chair as he talks, leaning forward, squaring his shoulders, and that his voice is getting louder and his words coming faster.

4. She doesn't allow her physical reactions to her emotion or Alex's behavior to obscure her cognitive ability to understand Alex's technical content. She hears that Alex is off track, taking a direction that goes further away from parameters that have a direct relation to motor thrust performance. He is not demonstrating the technically excellent problem solving approaches he usually is capable of using.

5. She tracks her sense that even though Alex appears confident he is also tapping his pencil, shifting his computer around, and not making sustained eye contact with his teammates.

6. She sees that her teammates are disengaged, they are looking out the window, at their notepads or laptops, anywhere but at Alex. Their emotional and attentional withdrawal does not match the highly technical content whizzing over their heads nor the time-sensitive aspects of the situation.

7. She realizes that anyone who was going to present an alternate solution to Alex has attempted to do so in previous meetings and will not try again. She takes a breath and adjusts her emotional response to be more in alignment with the technical content. An appropriate response on a team to potentially inaccurate technical approaches is to request teammates to find better solutions that actually work. *In other words, it would not be emotionally intelligent to avoid*

> confrontation with Alex and allow him to keep using valuable team time and resources trying solutions that won't work. Withdrawal and silencing herself at this point does not meet the technical needs of the engineering situation.
>
> 8. She squares her shoulders, straightens her spine, and chooses to focus on the confidence she has in the accuracy of her insight. She speaks up and interrupts Alex and offers her description of the alternate prediction model. Alex initially tanks right over her, insisting that he researched that model already and it is not applicable.
>
> 9. Several teammates sit up and take notice and start checking their notes for the updated predictions she emailed around last night. One speaks up and says Jaida's new numbers are closer to matching the results of the tests so far. Alex finally listens but not very enthusiastically, Jaida passes him her notes. After glancing at the team and noting their support of Jaida's determination, he calms down and focuses. The team agrees to try the alternate that addresses the updated parameters in the next test run in the lab. It works and their motor performs much closer to their updated predictions.

Jaida used microskills of emotional intelligence to balance cognition and emotion and positively impact engineering outcomes through this exchange.

EMOTIONAL INTELLIGENCE DEFINED

Emotional intelligence means we have the ability to balance reason and emotion.

Being emotionally intelligent means we balance affect (emotion) and cognition in our brain processing. We are neither cold, unapproachable geniuses like Mr. Spock, nor powerful but unstable emotional hurricanes like the Incredible Hulk. We are a mix of both. Here are some things you can do to be able to practice this balance.

Practice Behaviors that Express the Microskill of Emotional Intelligence

- Identify the emotions you are feeling in any given engineering setting.
- Track how your voice tone, body posture, facial expression, and word choice or listening skills are expressing these emotions.

- Track the voice tone, body posture, facial expression, word choice, and listening skills of others in your communication exchange.
- Listen carefully to interpersonal and technical content.
- Observe if emotional expressions in your voice tone, posture, facial expression, word choice, and listening skills match technical content. Do these expressions clarify content or obscure technical content reception between yourself and others?
- Adjust your own tone, posture, facial expression, word choice, and listening skills until your interpersonal and technical content are balanced.

For example, you may hear yourself getting impatient or sarcastic with a colleague who is nontechnical, for whom you are trying to clarify a technical aspect of a shared project goal.

You may see the confusion on your colleague's face increase rather than clear up as your impatience and frustration becomes evident on your face and in your voice tone and body posture.

You may know from past experience that when your colleague hears impatience and sees frustration on your face, your colleague's cognitive abilities for clarity get fogged by an anxious reaction, thus prolonging exchanges and bogging down task fulfillment.

Using your emotional intelligence to prevent this problem, you then check your posture, voice tone, facial expression and choice of words to adjust your technical content delivery for clarity rather than allowing your emotion (impatience) to obscure your technical content.

Using your emotional intelligence further, you also track your own discomfort so that if this person needs too many explanations just to get a simple task accomplished—you note your discomfort clearly enough that you self-motivate to find support from a coworker or supervisor to ensure you are not stuck in a position of clarifier at the cost of your own work time.

Try This

Consider how Jaida reached her self-understanding and acted with emotional intelligence even though she preferred to avoid conflict at all costs, especially in a group setting.

She tracked her own, measurable, physical, sensory responses. She observed those of others. She used this tracking and observation to support her cognitive choices about how to alter technical content in a team meeting. She balanced her emotion and her cognition and then made choices and took action based in that balance. This improved her own performance and that of the team.

You can do this too. Here's a start.

What is one of your favorite movies? What was fun or engaging about this movie? Did you enjoy the gut-wrenching thrill of adventure and near death? Did you enjoy a complex plot driven by intense characters that pulled you into their lives?

Make a list of the emotions you experienced while watching your favorite movie. Recognizing scenarios that engage your imagination, please you, entertain you, make you laugh or cry can be a first step in gaining self-understanding and emotional intelligence.

In your favorite movie, what language choices, posture, facial expressions, voice tone, and pacing of interaction did the characters use to convince each other of important information or personal needs? How did they get each other to hear accurate details? Did their words and meaning sometimes get lost in the shared interpersonal field of interaction? How did that "cloudy weather" impact plot outcomes? How did they clear through this cloudy weather and choose actions that made sense or had productive outcomes?

Jaida's experience and your favorite movie probably have something in common. They are stories of characters dealing with pressure and choices. They are stories of characters managing to not be overwhelmed by their own or others' emotions in the process of either making a capstone project or saving the world. But they are also stories of characters who had emotion, who observed this emotion and used it as a resource rather than a dangerous force to be excluded from a situation. This is important because emotion is a "physical" force, and like gravity or voltage in an engineering equation can lead to design failures if not properly taken into consideration.

CHECK IN

Did you ever have a teacher, coach, or family member who mentored you in a way that you experienced as positive and helpful?

How did that person talk to you, focus attention on you or give you some reflection about your own professional or personal growth? How did being with that person feel to you?

Chances are good that the person who had positive impact for you was emotionally intelligent in that area of their lives. They were able to stabilize their own inner process enough to give you the kind of attention you needed to grow.

CONTEXTUAL INTEGRATION OF EMOTIONAL INTELLIGENCE

Emotional intelligence involves a capacity to adapt brain-based, information processing of cognitive responses to include sensory affective (emotional) responses. It drives self-understanding and modulates appropriate affective processing.

In an article on successful team function in the Harvard Business Review, "Building the Emotional Intelligence of Groups," Vanessa Urch Druskat and Steven B. Wolff, describe Daniel Goleman's seminal clarification of this essential human capacity,

> In his definitive book, Emotional Intelligence (EI), Goleman explains the chief characteristics of someone with high EI; he or she is aware of emotions and able to regulate them—and this awareness and regulation are directed both inward, to one's self, and outward to others.
>
> —Druskat and Wolff (2008)

Put in behavioral terms that lead to microskills usage, emotional intelligence is practiced as the skill of tracking the physiology of emotion during communication exchanges and ensuring that our emotions stay balanced. Ensuring that our emotions do not either wash over or become disconnected from our cognitive choices.

We want to emphasize the word "balanced" here. Balanced emotion and cognition means that emotions are not pushed out of the picture to the point of being ignored nor are they allowed to impact a design decision or team discussion so fully that they overwhelm clear cognition.

In achieving this balance, we use tracking of body changes and observation of others' responses to allow our emotions to provide rich and vital motivations that inform our cognitive choices toward productive interaction and endeavor.

You will be learning to monitor your own self-understanding and emotional intelligence through the use of self-modulation in clear affective processing. Engineering proficiency in both technical and interpersonal modes requires this type of monitoring from you toward yourself and others to become robust.

CHAPTER 6

AFFECT CHANGES YOUR COMMUNICATION

Affect leads to the microskill tab of affect processing in our Communication Microskills Model. Affect strongly bonds other microskills into combinations that can clarify or obscure interpersonal and technical engineering communication exchanges.

> Experiences, thoughts, actions, and emotions actually change the structure of our brains.
>
> —Ratey (2002)

You probably have a clear understanding of the meaning of the following words from the quote above—experiences, thoughts, and actions. You probably also have some idea of how these can impact your brain development and your communication needs as an engineer.

But what about that word "emotion"? How does it impact the comprehension of technical and nontechnical interpersonal communication in engineering? Emotions (or feelings) impact the comprehension of technical and nontechnical interpersonal communication in engineering because affective processing (emotional responding) is a primary developmental foundation of all brain processes

Effective Interpersonal and Team Communication Skills for Engineers, by Clifford A. Whitcomb and Leslie E. Whitcomb.
© 2013 by The Institute of Electrical and Electronics Engineers, Inc. Published by 2013 John Wiley & Sons, Inc.

- As infants we learn emotion first, then primary language anchored in gut-based emotion, then abstract language and reasoning informed by complex emotion.
- This sequence happens every time we hear and respond, even as adults.

Every piece of information we receive in a technical communication goes through our emotional processors before it is integrated by our cognitive processors.

Note: If you read nothing else in this book—read the phrase above, re-read it, and find a way to understand it's impact.

The following dialog shows you how this works in engineering communication settings.

A Dialog Demonstrating the Impact of Affect on Technical and Interpersonal Engineering Communication Contexts

At 3 o'clock, Dex and Michelle's interrupted conversation about the Smartphone app can be completed. Reading Dex's description of overcoming the programming glitch that was driving him nuts gives you a good idea of how his affect was a primary experience that he eventually mediated with cognition and behavior to achieve a step forward in technical engineering tasks.

Dex and Michelle are now meeting in the design lab. They have their laptops open. Dex is showing Michelle the data query code sequencing he finally found that worked, after hours of struggling.

Dex, "See—here is where you got the merge of tweets, campsite maps, and activity suggestions to work. I was in there last night trying to see every step of the code you used so I could get the pictures in there too."

Michelle, looking at the pictures they're trying to include, "Wow, remember the drop from that rock ledge in the canyon? When your water bottle went over the side it was pinging all the way down—all I could think right then, was what if that was somebody's phone? But yeah, there are some differences with images, but that still didn't explain to me why it was so hard to figure this one out, after all, if the thing as a whole is pretty much working, why would one aspect go nutsoid on us?"

Dex, "Yes, exactly. That's what kept me at it for hours last night. I couldn't stand it that only one aspect of the thing—Just. Wouldn't. Work. When the others ones did!"

(Affect—emotion—is now present in the dialog. It is no longer a purely cognitive based, technical exchange. Dex is upping his voice volume and adding exclamation points to his content delivery. It is clear the frustration was strong enough to keep him engaged with the problem into the wee hours.)

Michelle, smiling and looking up from the screen, "Yuh Huh, that's you all over Dex. You will not give up—never say die. You went to the top of that mountain on the canyon hike even though the rest of us were ready to quit and back it down. But sometimes you shoot yourself in the foot, dude. Overdoing it doesn't always get something solved."

Dex, smiles and shrugs, "You got me. Were you floating around the ether in my apartment last night or what? I was like that guy, Jerry Seinfeld, in the animated Bee Movie. Where he bumps into a closed window, trying to get out, and then just keeps hitting his head, saying— 'Maybe this time, maybe this time, this time, this time . . . ,' until he's dizzy and his head hurts but obviously the window is still there, and he then he says '*diabolical.*'"

(Michelle laughs appreciatively, Dex continues)

"But seriously. I kept repeating the coding like we did them for the other aspects of the merge. Over and over and over again. I kept thinking I saw it right, then I'd try to get the pictures right in there with the campsite. And I'd press the button and they wouldn't be there! The madder I got, the more times I repeated it, the more times it wouldn't work."

(Dex is describing a moment in which his primary response is affect based—the desire to get it right over ruled his capacity to think of and then choose another approach. He was having the standard, affect-first response, we all have, but not able to take the affect moment—the frustration—and move past it into cognition, a new idea, and then into proactive behavior.)

Michelle, laughing, "So was the magic of sheer repetition what solved your problem? Which by the way if it was, you'd be the first person in the history of programing to get it that way. (Dex laughs and shifts the screen view.)

Michelle continues, "But seriously, Dex, how'd you figure it out?"

Dex, "Well, remember I have one of those ring tones to tell me a sports score? The thing toned and interrupted me and I took a break to read the score. I got something to eat, came back to the computer. And then guess what! I realized the first time I looked at it that I was doing the query too early. Then, and only then, did I remember that at the end of the lab session yesterday you told me that especially with images the query needs to be treated differently.

In my frog-face enthusiasm to get it right, I just did not remember that. Once I adjusted the query, it all fell into place. Then I was so pumped I took it one step further and programmed in some swipe gestures so you can switch easily from viewing a map to viewing the campsite and a picture of the surrounding area at the same time."

(Dex got his affect, cognition and behavior sequencing in balance after taking a break. At first, his affect had blocked the clear technical content Michelle had given him at the end of their lab session. After taking a break and clearing his affect, he was able to remember the technical information and then to use the drive of his affect to create technical clarity rather than create obstacles to a solution.)

Michelle, "Nice work, Dex. Let's see if we can polish it off and then send everyone a tweet that it's available and have them try it out for us."

The accepted scientific term for the experience of emotion that Dex went through is *affect*. Affect is researched and understood as a key function of how you process internal and environmental stimuli, including technical data, in your daily life.

AFFECT DEFINED

Affect means a feeling or the experience of emotion.

What this has to do with communication in engineering situations is that your brain/body responds to the stimuli of engineering technical and interpersonal content in your communications with an affective reaction.

This reaction has to happen and be balanced first or your brain never gets to accurate perception, association, and resultant cognition.

This reaction cannot be ignored—as Dex did when he repeated his task too long even though it wasn't working.

And it has to happen in a way that doesn't allow affect to obscure cognition. As it did when Dex forgot key information from Michelle that was technically relevant to his problem.

To avoid these types of engineering task flow blockages, it is important to track your affect, cognition, and behavior sequencing in any given engineering communication exchange.

- Affect happens first and happens very quickly—even before our cognition can catch our reactions.
- Cognition happens second and is based on affective response.
- Behavior happens last in the sequence and is based on our capacity to process affect into constructive cognitive choices for behavior.

AFFECT DEFINED IN EVERYDAY LANGUAGE

Our culture uses short simple words to describe affect.

Everyday Emotions that Impact Technical Engineering Communications

Primary Feelings: sad, mad, happy, surprised, confident, frustrated, frightened

Relational Feelings: engaged, withdrawn, attracted, disgusted, attached, alienated

Social Feelings: concerned, empathetic, defensive, assertive, passive, proud, embarrassed

---

This affect expressed as emotion shifts to the background during technical communications but as we saw in the example with Dex and Michelle, it never completely disappears from the brains and bodies of engineers while they are negotiating technical issues. *That is because each cognitive function you have originates in everyday emotions. Our brain processes most stimuli, including cognitive stimuli, through affective processors first and then subsequently through cognitive brain function centers.*

CHECK IN

How do you know when you are having a negative experience in an engineering team meeting? Does your thought process tell you that things are not going well solely through the medium of language? Or does your body record stress? If so, how?

How do you feel when your project deadline is suddenly moved up 2 weeks due to scheduling constraints? Are you relaxed in your chair? At ease? Smiling and laughing and feeling like you just won a grant or scholarship? Or do you feel differently?

Did you ever make a mistake in calculations or design projections when you were tired, stressed, just broke up with someone, or felt your position on a team or on a job was at risk?

What cues do you follow to let yourself know that you have alienated a teammate? Are you consistently surprised by this information because you didn't see it coming? Or have you noticed cues coming and simply felt they were unrelated to technical content and thus unimportant unless they caused design and costing mistakes? This happens frequently in engineering settings. Learning the microskills of affect can prevent it.

CONTEXTUAL INTEGRATION OF AFFECT

Your affect is tracked in your brain through physical responses that trigger the perceptual and associative memories on which you base your cognitive and technical communication content choices.

This doesn't mean you are not a technical genius or a proficient engineer. It simply means you are human. It also means that if you are under creative stress or stress induced by technical constraint management or economic/ work based stress, *your brain balance of emotion and cognition can be altered because the neurophysiological volume of your emotions may cloud the clarity of your cognition.*

When you experience high creativity, or even mild stressors, your body records emotional processes triggered in your brain from these environmental stimuli. You then respond with changes in breath rate, heart rate, brain pulse, body posture, muscle tone, facial expression, hearing acuity, visual acuity and facial expression. This takes time, and some brain focus, away from primarily cognitive processes. In the emotional intelligence example, Jaida could have allowed the emotional volume of her shyness and need for withdrawal to cloud her cognitive understanding that new technical solutions were necessary for the future success of her rocket launch project.

Just like Jaida, you may find that when you are at ease, an adequate balance is maintained between tagging emotion and practicing cognition, *but when you are stressed or highly creatively engaged, an imbalance can occur.*

The need to balance your emotional and cognitive responses even while experiencing technical and interpersonal stress is why affective processing becomes a key function of engineering proficiency.

Jaida balanced her emotion to support her cognitive responses, but she didn't do so by ignoring her affective responses. She did so by noting and anchoring herself in the physiological sensing of those responses to support the cognitive clarity she needed to create proactive engineering outcomes.

CHAPTER 7

AFFECT PROCESSING: THE HIDDEN KEY TO CLEAR COMMUNICATION

Affect Processing is the microskill tab in our Communication Microskills Model that is always there, impacting all the microskills during communication. By being aware of the effects of affect processing, you can become the master of your communications in all situations.

> The result is that everything you do has both a cognitive and an affective component—cognitive to assign meaning and affective to assign value. You cannot escape affect: it is always there. More important, the affective state, whether positive or negative affect, changes how we think.
>
> —Norman (2004)

Affect processing helps you work with affect even though you cannot see it, taste it, smell it, or measure it. Microskills of *internal affect processing*, *external affect processing*, and *balanced affect processing* will give you the basic selfunderstanding and emotional intelligence to master all the skills in this book. These skills address the reality that

- You process affect by understanding the tags it leaves in your own body—and in the facial, vocal, or gestural behaviors that serve as indicators of affect in others as you communicate with them.
- Affect, like gravity, is a force that can best be worked with and understood through its impact on you and your environment.

Effective Interpersonal and Team Communication Skills for Engineers, by Clifford A. Whitcomb and Leslie E. Whitcomb.
© 2013 by The Institute of Electrical and Electronics Engineers, Inc. Published by 2013 John Wiley & Sons, Inc.

To help you make the leap into working with affect, we first offer an engineering analogy.

An Electrical Engineering Processor Circuit Analogy

A processor is a multifunctional device that at the most fundamental level transforms inputs into outputs. An audio amplifier can designed using a processor, taking in signals by receiving low amplitude input signal and transforming it into a high amplitude output one. Ideally for a linear amplifier, the fidelity of the signal entering matches the fidelity of the signal leaving. The amplifier gain provides an adjustment for the amount of amplification of the input signal. Too much input, or setting the gain too high, can overdrive the circuit. Overdriving pushes the output beyond design limits, so it is no longer linearly proportional to the input—leading to severe distortion and loss of useful output information. Too little signal will not provide enough strength to overcome any input bias. Opening the path to or from the circuit will cut off the amplification function, leaving it inoperable. For either the low input signal strength or open circuit case, there is no output from the amplifier that contains useful information, since it is no longer related to the input.

INTERNAL AFFECT PROCESSING DEFINED

Your internal affect processor is neurophysiologically designed to tune into *sensory input* cues as signals for processing your heart rate, muscle tone, breath rate, perspiration, posture, and overall affective tone. Your body then uses this neurophysiological information to form, modulate and transmit affective outputs as you create, think, and communicate. Affective outputs are expressed through heart rate, body posture, perspiration, breath rate, muscle tone, voice tone and frequency, and facial expression.

Sensing, tagging, and responding to internal cues, instead of ignoring them or reacting against them, is internal affect processing.

EXTERNAL AFFECT PROCESSING DEFINED

Your external affect processor is neurophysiologically and culturally designed to tune it's inputs into *sensory output* signals that others transmit as cues—their posture, facial expression, voice tone, gestures, semantics, and technical content—the overall affective tone of their communications to you. These cues enter your body and set off internal affect processing that will drive and influence both technical and interpersonal responses you experience.

Sensing, tagging, and responding to affective cues in others' technical and nontechnical interpersonal communications, instead of ignoring them or reacting against them, is external affect processing.

BALANCED AFFECT PROCESSING DEFINED

Your affect processors—both internal and external—allow you to adjust the gain of input and output so affect does not overdrive or disconnect your communication content.

Balancing your response to internal and external affect processing creates clear and accurate engineering communication.

Try This

Make a list of emotions that one of your favorite movies elicited in you. Trace the flow in the circuitry of that emotion in your memory. Did you enjoy the movie because it flooded you with curiosity, suspense, or laughter? Did you enjoy the movie because it sent you into a world of imagination that allowed you to totally tune out your daily stresses? When you were hooked on adventure did your muscles relax into total focus on the experience in front of you? When your favorite character was about to lose their life did you tense up, hunch your shoulders, hold your breath? When the plot line became more complex and difficult to predict did your eyes move rapidly back and forth and did your breath become shallower as you tried to figure it out?

These responses are cues you can follow to begin tracing how you respond in stressful and creative engineering exchanges. When you begin to trace these cues you are beginning to learn your emotional palette. Your emotional palette includes all the primary emotions in the box at the beginning of this section and it colors all of your technical and nontechnical engineering communications. When you can recognize these primary emotions fluently in many different situations and balance them so they do not cloud your cognitive choices. You are learning to equilibrate internal affective processing with external affective processing to have balanced affective processing. Then you are headed toward being proficient in clear affective processing. This is important because it helps you avoid overdriven and disconnected affect responses.

AFFECT PROCESSING IN OVERDRIVEN, DISCONNECTED, AND CLEAR RANGES

Every microskill you will learn in this book is based on your ability to modulate internal and external affective responses toward balanced affective responses. This is *Clear Affective Processing*. Clear affective processing is your ability to modulate internal and external affect sensing and expression. It is easier to understand and gain this skill by first understanding and observing interactions in which this skill is absent. The following demonstrates *Overdriven Affect Processing* and *Disconnected Affect Processing*.

Overdriven Affect Response

A Practical Example of Overdriven Affect Processing in an Engineering Setting

You openly gave a teammate heat for not delivering a project deliverable on time. Then you walk away, and after having some downtime, you realize that your own time constraints and workload actually created an obstacle to your teammate completing their deliverable. You allowed your fear of failure and frustration with the project to overdrive your affect processor. This distorted your capacity to intentionally control affect to *serve* function, rather than *override* function, and your communication clarity was compromised. Your fear and frustration turned into an inaccurate situation assessment, and you blamed your teammate instead of taking the blame yourself.

Overdriven Affect Processing Defined

Overdriven Processor Response—occurs when you are sad, mad, glad, afraid, surprised, frustrated, and/or experiencing variations of these primary emotions (as in our list that names basic human emotions).

Overdriven Affect Processing Analogy

Overdriven affect processing occurs when the output is amplified as you experience these emotions to the point that your heart rate, breath rate, perspiration, muscle tone and thoughts lose equilibrium, and selfmodulation thresholds are crossed. Your affect processor is receiving too much input, the circuitry is at risk of getting overloaded or disconnected, increasing the potential for unintended output (miscues, conflict, self-silencing, or alienation of teammates and co-workers).

Overdriven Affect Processing Impacts on Engineering Communication

Intentional affect control allows you to avoid overdriving your affect processor. You are then more capable of achieving clear affect output. When you are not proficient at receiving your own signals and controlling your own processing, your affect processor is at risk of going into overdrive on reception, and this necessarily impacts output. An overdriven processor response occurs when your amplified reaction to being sad, mad, glad, afraid, surprised, or frustrated redirects your own affect into misplaced action or blame.

Internally: This translates into distorting the sensing or transmission of your own affect cues—reacting rather than responding to the fact that you are sad, mad, glad, afraid, surprised, or frustrated.

Interpersonally: This translates into distorting the reception of others affect processor amplification cues, or unintentionally modulating your voice tone, gestures or postures in such a way that the functional content of your communication is lost in the delivery.

Disconnected Affect Response

A Practical Example of Disconnected Affect Processing in an Engineering Setting

You are assigned to a team role that does not match your proficiencies. You are not comfortable feeling or expressing the sad, mad, afraid, and frustrated sensor signals your affect processor sends you at the moment you are assigned to this role. You ignore the tight breath, stiffening of posture, racing thoughts, and sinking feeling in your gut. You assume your team leader cannot change your role or your teammates will be so assertive in keeping their preferred roles that it might not seem worth the effort to register and express your affective distress cues, *even to yourself in your own perceptions*. Your teammates mistake your signals as acceptance of your role and do not give you the attention that someone exhibiting more obvious signs of distress might receive.

You take it on the chin, focus on the technical aspects of your role and soldier on. In some situations this may be appropriate and necessary. But if your role is truly not compatible with your proficiencies it can lead to a poor project outcome, a note on your record about your less than excellent performance, and an eventual blockage of your move to a better matched team or a more successful project. Conversely, if you do so well at masking your distress that you actually shine in your incompatible task, you may find yourself consistently pushed in a career function and direction that has nothing to do with your preferred career or personal desires or intentions—and this can lead to low personal satisfaction.

Disconnected Affect Response Defined

Disconnected affect response occurs when you are sad, mad, glad, afraid, surprised, frustrated, and/or experiencing variations of these primary emotions (as in our list that names basic human emotions).

Disconnected processor response occurs when you experience these emotions to the point that your heart rate, breath rate, perspiration, muscle tone, and thoughts lose equilibrium and selfregulation limits are crossed. You have had a sensor input overload and respond by disconnecting.

Disconnected Affect Processing Analogy

Disconnected affect processing is like opening an electrical circuit. Opening a circuit compromises its ability to transfer signals. Cutting off the input opens the circuit completely. The circuit has been defeated—it has been effectively disconnected from the input signal is designed to process. Now it is not providing enough input signal to overcome any input bias also prevents a circuit from functioning. There is no longer enough signal input to overcome the bias to trigger the output. *Although those input signals are still present in the environment, the processor is not receiving, and subsequently not outputting any useful signals.*

Disconnected Affect Processing Impacts on Engineering Communications

A disconnected processor response occurs when your reaction to emotion becomes an internal assessment that the person, situation, or dynamic you are dealing with is an unstoppable force meeting an immovable object—in reaction to this assessment—you shut off your circuit that processes your affect. This bears repeating, you shut off your circuit that processes your affect.

When you make an unintentional assumption that your affect and responses will not change or fix the situation—you have cut off your affect processor—and thus you default to practical or technical solutions provided by the only processor left connected—your cognitive process. You are then unintentionally modulating the amplification of your presence and responses without the benefit of your affect processor—no longer processing feelings and relationships—just processing the facts.

Internally: This translates into lowering the volume of affect sensing of breath rate, heart rate, muscle tone, perspiration, posture, and overall response tone inside yourself to the point that there is not enough signal on the input to overcome the input bias—to trigger your affect processor. You lack selfawareness, so you are not accurately perceiving yourself in the situation.

Interpersonally: This translates into losing attention and focus thus lowering the output of the affect cues you transmit through voice tone, gesture, posture, and overall body tone to the point that you have defeated the circuit for communication. When your cues are lowered in this way you are also simultaneously lowering your capacity to transmit your intended cues and to process the cues of others. You are participating in the communication exchange—but you are neither present enough for others to receive your intended affect signals, nor are you present enough to receive the cues of others.

Clear Affect Response

Dylan, Don, Lin, and Jaida all demonstrated clear affective processing in the examples we have given you thus far in the book. Dex demonstrated over-driven affect processing that moved into a clear affect response.

Clear Affect Response Defined

Clear Affective Processing can occur when you are sad, mad, glad, afraid, surprised or frustrated, and/or experiencing variations of these primary emotions as listed in our initial box naming basic human emotions.

When you are able to sense your own affect and *respond* to your affect rather than *react* to your affect, then you can intentionally control the regulation of your affect.

Clear Affect Response Analogy

Clear affective processing occurs when the fidelity of the signals received and sensed internally match the fidelity of your affective output.

Clear Affect Response Impacts on Engineering Communication

This control allows content of communication—both interpersonal and technical—to be received and transmitted with affect that adds emphasis, is persuasive, is inviting and positive and that engages others' enthusiasm and cooperation.

Internally: This translates into the ability to use your affect processor sensing to read the data from your breath rate, heart rate, muscle tone, posture, and overall affective tone that are input as signals to you. It translates into the ability to amplify sadness, anger, fear, excitement, surprise, and frustration as informative emotions rather than reactions that overdrive or disconnect communications.

Interpersonally: This translates into the ability to amplify disappointment, anger, fear, enthusiasm, surprise, and frustration with *intentional* controls. Intentional control means *you can modulate* your affect with voice tone, body posture, gestures, semantics, and overall tone along with the technical content to convey holistic communication—communication that is vibrant and supports your roles and functions as well as that of others.

Try This

Make a list of emotions that one of your favorite movies elicited in you. Trace the flow in the circuitry of that emotion in your memory. Did you enjoy the movie because it flooded you with curiosity, suspense, or laughter? Did you enjoy the movie because it sent you into a world of imagination that allowed you to totally tune out your daily stresses? When you were hooked on adventure did your muscles relax into total focus on the experience in front of you? When your favorite character was about to lose their life did you tense up, hunch your shoulders, and hold your breath? When the plot line became more complex and difficult to predict did your eyes move rapidly back and forth and did your breath become shallower as you tried to figure it out?

These responses are cues you can follow to begin tracing how you respond in stressful and creative engineering exchanges. When you begin to trace these cues you are beginning to learn your emotional palette. Your emotional palette includes all the primary emotions in the box at the beginning of this section and it colors all of your technical and nontechnical engineering communications. When you can recognize these primary emotions fluently in many different situations and balance them so they do not cloud your cognitive choices. You are learning to equilibrate internal affective processing with external affective processing to have balanced affective processing. Then you are headed toward being proficient in clear affective processing.

CONTEXTUAL INTEGRATION OF AFFECT PROCESSING

Why is it necessary to practice internal, external, and balanced affective processing in engineering communication exchanges?

- Because *affect is a given* in any interpersonal communication situation we encounter. Affect drives 87% of our technical and nontechnical exchanges—since this is the percentage that is *nonverbal* (Mehrabian, 2007; Ekman, 2007; Knapp and Hall, 2007). Significantly, nonverbal communication is experienced and processed in areas of the brain that function primarily from a basis of affective cue sensing (Greenspan, 2001; Tronick, 2008; Ratey, 2002). *This is a highly important communication strategy for you to know because it means that the nonverbal stimuli of communication, 87% of all communication—technical and nontechnical—happens in brain centers that are strong in affect and weak in cognition.*
- Because *affect is present* in any interpersonal communication situation *regardless* of whether or not your processor is receiving from your own or others' affect signals.
- Because when you are not proficient at sensing the signals of your own affect cues, or receiving the signals of others' affect cues, then the gain on your affect processor is controlled by others, the situation, or your own unintentionally triggered and unintentionally amplified affective feedback.
- Because when the fidelity of the affect signal received does not match the fidelity of the affect signal transmitted, then you have an emotional disconnect in the communication exchange. This can defeat the entire exchange, and lead to a breakdown in the encounter. A breakdown in an engineering communication exchange means a breakdown in the engineering processes of conceive, design, implement and operate. So clear affective processing leads to successful engineering communication exchanges that lead to successful engineering outcomes.

Now that you have a beginning sense of the Space, Face, and Place Spectrum on which you communicate, you have identified your own natural style of communicating in engineering settings and you are aware that your communication exchanges need to be balanced interpersonally and technically—cognitively and emotionally. You are ready for the nuts and bolts basics of actual microskills usage. We begin with "I," "You," and "Team" statements in engineering communication exchanges.

SECTION II

TAKING IT TO WORK

Initial microskills learned in this section involve the proper use of content statements that clarify communication. In the section as a whole you will learn consideration of how you attend to your communications (including both verbal and nonverbal aspects) you will learn how to pose open and closed questions as part of your dialogue, and you will learn to combine all of this together in a multimodal, interpersonal and technical, sense. The first skills to learn are those of I, You, and Team statements.

Effective Interpersonal and Team Communication Skills for Engineers, by Clifford A. Whitcomb and Leslie E. Whitcomb.
© 2013 by The Institute of Electrical and Electronics Engineers, Inc. Published by 2013 John Wiley & Sons, Inc.

CHAPTER 8

I, YOU, AND THE TEAM

I, You, and Team statements are microskill tabs in our Communication Microskills Model that serve as invitations to use more dynamic, resolution focused information exchanges in interpersonal and technical engineering situations.

The need for communication effectiveness with clear "I," "You," and globalized "Team" statements is not an abstract concept. The building blocks of good communication consist of appropriate "I," "You," and "Team" statements.

Good Communication—"I" Statements

- I feel
- I think
- I believe
- In my opinion
- I'm frustrated
- I'm angry
- I can work with that
- I'm okay with that.

Effective Interpersonal and Team Communication Skills for Engineers, by Clifford A. Whitcomb and Leslie E. Whitcomb.
© 2013 by The Institute of Electrical and Electronics Engineers, Inc. Published by 2013 John Wiley & Sons, Inc.

"I" STATEMENTS DEFINED

"I" statements generally use the word "I" to express your perspectives, needs, statements, and views.

- focus on your own thoughts, feelings, and actions,
- focus on what you know, your thoughts, and feelings,
- make a request for help or change.

Try This

For each communication exchange below, think about how you might restate the given statement to make it an "I" statement.

Choose one communication exchange below and consider how to tag it with a single word, sense based feeling descriptor, that is, "tight chest," "head pressure," and "sigh of relief."

1. "You're not contributing to this project enough."
2. "We're not getting anywhere."
3. "You don't care if we succeed."
4. "You won't let anybody else at this table get a word in here."

For the following situation, think about how you might prepare an "I" statement to express yourself.

5. You did not complete the action items that you signed up for at the previous meeting.
6. You and some other teammates agree that another team member takes over the whole conversation in the meetings.
7. You are in a meeting and the team keeps going off topic.

Once you have mastered a capability to make your communication happen from the center of your own communication intent with "I" statements, you can add functionally placed "You," and "Team" statements to invite others into a communication dynamic with you that moves task fulfillment forward for your self, your peers, and your team as a whole.

CHECK IN

How do you feel while reading "I" statements?

How would you feel as the recipient of the "I" statements?

"I" statements are the functional expression of self-understanding and emotional intelligence. They are structures that support a balance of emotion and reason in communication exchanges. Are you comfortable making "I" statements? Or would you rather express yourself with the buffers of technical needs, group needs, and referencing the actions of others as a way to describe your own re-actions?

CONTEXTUAL INTEGRATION OF "I" STATEMENTS

Using "I" statements says that you take responsibility for what you are saying, seeing, feeling, and doing. Actions taken based on "I" statements help meet expectations for getting things done, since everyone has stated what they have agreed to do in their own terms, and in the presence of other team members. "I" statements can be seen as the engine of not only coordinated team communication but also coordinated team action.

Learning the use of "I" statements might seem like an almost kindergarten-level of basic communication. It is, however, the fundamental nature of these statements that gives them the power to either drive task fulfillment or inhibit progress, because "I" centered statements are so basic to human communication. Their effective usage can transform illogical responses and functional errors into clear thinking and proactive task completion.

To better understand the fundamental nature of "I" statements, try to think about any communication exchange you had over the past twenty-four hours. Your exchanges most likely included situations where "I" statements were necessary. When greeting a friend, coworker, or colleague, you might have said, "Sorry I'm late," or "I'll be right with you" or "Sorry I missed yesterday's meeting, can you please fill me in?" Notice that in each of these statements, you are taking responsibility for what you know about yourself, what you are requesting and what shared actions can go forward from the exchange.

Even better, "I" statements allow you to provide your teammates with information about your intended content and your perspective on a given situation at the same time. The following "I" statement from a team meeting illustrates this concept.

"I understand the analysis results in this discussion, but I don't see how the dialogue is moving us forward."

By using this statement you are not only expressing that you are engaged in the meeting, but you are expressing something only you can know (that you understand the technical content) and focusing attention on the specific concern you have. You are expressing YOUR concern that the discussion is not going to move the project ahead. This leaves room for your team members to respond knowing that you need to be convinced that they are headed in the right direction, or that you might be correct, and everyone needs to reconsider the context for the discussion. In engineering terms, this is similar to an elegant solution, a seemingly simple solution that meets multiple needs simultaneously. This solution cannot happen if you are using *Opaque "I" Statements*.

OPAQUE "I" STATEMENTS: AN EXAMPLE

The following is an opaque "I" statement made in a team meeting.

"Your numbers make sense but you're not dealing with them correctly," or, "This discussion is getting us nowhere."

This expresses the same communication as the example given above in the "I" statement section, but it does so with roadblocks and confusion rather than an invitation to move forward in team task fulfillment.

OPAQUE "I" STATEMENTS DEFINED

Opaque "I" statements are those that replace the "I" with a "You" or "We" (in the context of the team or perhaps more generally in a global sense—like the "royal We") word—what we call misplaced "You" and "Team" statements'. They are opaque because they hide your perspective behind a facade of another's.

CONTEXTUAL INTEGRATION OF OPAQUE "I" STATEMENTS

When "I" statements are substituted with misplaced "You" and "Team" statements, then communication effectiveness is compromised. Your team members now do not have a grounded context within which to understand what you might really be saying. They have to guess at what you mean. Is something wrong with what is being discussed? Is there an unspoken problem? Is one of them individually messing up? Or is it that the team is not functioning well? When you desire an action or change or have a need to contribute your perspective, using an "I" statement makes the intention of your communication clear and concise, and avoids having everyone fill in the blanks—with possible bad assumptions that divert creative energy into the discussion rather than into task completion.

Using an opaque "I" statement, wrapping your intent and perspective in "You" and "Global" statements, obscures your meaning. Worse, it focuses the implied initiative for action or change *on your peers only* or *the group as a whole*, leaving *you and your perspective* out of the dynamic.

This can happen when "You" and "Global" statements are used in the following ways.

MISPLACED "YOU" STATEMENTS: EXAMPLES

- You always
- You never
- You think
- You don't care
- You can't
- You won't.

MISPLACED "YOU" STATEMENTS DEFINED

Misplaced "You" statements are those that

- focus on what you do not know, the thoughts, feelings, and actions of others,
- ask the other person to act or change,
- can feel like blaming.

CONTEXTUAL INTEGRATION OF MISPLACED "YOU" STATEMENTS

You as an individual person cannot know what another person is thinking or feeling. So using misplaced "You" statements to express how *you* yourself are experiencing a situation is not effective communication.

Additionally, asking another person to act or change as a way of expressing your own experience or needs is not effective communication. Another person's behavior can never really be an accurate reflection of your internal feelings and perceptions. So expressing yourself and your perspective through "You" statements can feel directive and nonnegotiable to the person receiving your information.

Finally, when you communicate by using "You" statements instead of expressing your own thoughts, feelings, or actions—then your unexpressed emotions or thoughts and frustrated actions can add an edge to your communication that might feel like blaming and criticism to others. They are more easily received as a closed door, a noninvitational position.

Then the people you are communicating with might feel a need to spend time and energy defending their own thoughts, feelings, and actions rather than putting their energy into working together on the tasks at hand. Misplaced "You" statements are often made even more complicated by being delivered right along with misplaced "Team" statements.

MISPLACED TEAM STATEMENTS

- We're missing the point.
- They never get anything done on time.
- We never get anywhere in these team meetings.

MISPLACED "TEAM" STATEMENTS DEFINED

Misplaced "Team" Statements

- Make general descriptors of group and team realities that might not be shared by all.
- Are stated as if they are inarguable facts.
- Can sound as if things cannot be changed or negotiated.

CONTEXTUAL INTEGRATION OF MISPLACED "TEAM" STATEMENTS

When you do not actually want to individually criticize or blame but you still are attempting to express your own experience without making "I" statements, it is easy to fall into making misplaced "Team" statements.

These are statements that make a general description of a situation but do not really address your place, needs, actions, or perception of the situation.

This misplacement invites forms of expression that are globally based "Team" statements.

Because they are floating around out there in the ether of shared communication they are not grounded in what you can definitely know—your own thoughts, feelings, and perspective. Nor are they grounded in a consensus based group choice about a perspective. You may end up claiming "fact," "norms," "everybody does it this way," positions to anchor your thoughts, feelings, and experience.

When your misplaced "Team" statements are ungrounded and thus need to be explained by using more global, blanket statements, they often can sound nonnegotiable or resistant to change.

Because you communicate in groups as well as communicating as an individual, misplaced "You" and "Team" statements cannot always be remedied by using appropriate "I" statements. You need to also know how to make constructive "You" and "We" statements to make your communications clear and proactive.

YOU AND WE COMPLETE THE EXCHANGE

"You" statements clarify your communication exchanges when you task peers and teammates, give information about the performance or roles of co-workers and peers, request information from advisors and supervisors and elicit needs of stakeholders and consumers.

Good Communication—"You" Statements—Examples

- "Anna, can you please give me that analysis before the meeting winds down."
- "you can create the customer surveys"
- "would you mind repeating the tasks you assigned to me?"

APPROPRIATE "YOU" STATEMENTS DEFINED

"You" statements clarify your communication exchanges when you task peers and teammates, give information about the performance or roles of co-workers and peers, request information from advisors and supervisors and elicit needs of stakeholders and consumers.

Because teams communicate through individual team members talking to each other and also as a cohesive set of individuals with a shared understanding that their statements are expressed based on a unified consensus, appropriate "We" statements are also a necessary component of effective engineering communication.

GOOD COMMUNICATION: "WE" STATEMENTS—EXAMPLES

- "We did it!"
- "We can schedule the next meeting before we break up."
- "We have seven days left until the prototype demonstration."

APPROPRIATE "WE" STATEMENTS DEFINED

Using "We" takes the place of "I" in team communication statements, showing that the team takes responsibility for the communication. We call these statements "Team" statements, statements that you use to clarify your communication exchanges when you are interacting from a team context.

In real life situations "I," "You," and "Team" statements are used interchangeably and fluently within even very brief communication exchanges. Putting it all together makes the exchange a balanced, effective communication.

Try This

For each of the following situations, prepare an, "I," "You," or "Team" statement that is appropriate for the context and intent of the communication.

1. You are in a design meeting and the team has strayed from the agenda and nothing is being accomplished.
2. You are the team leader and you have to assign roles for initial design responsibilities.

I, You, the Team and Affect

Now add affect tags to the "I," "You," or "Team" statements you crafted above.

First, name a sensory (affect tag) response you would have in the stagnant team situation. For example, in the team meeting where nothing is being accomplished you may be a person that would feel bored right out of their socks. How do you know you are bored? You know it because you think it. But you also know it because you feel it. What does your body feel like when you are bored? Numb? Sleepy? Slumped? Edgy? Distracted?

Second, name a situational shift (environmental stimulus) that would re-engage your interest. For example, if someone introduces an idea that will mean lots of extra after-hours work for your contribution. You are going to go from boredom to being on the alert pretty quickly. How does this shift feel? Like a cold shower wake-up call? Like a mini electric jolt up your spine? Like a motor shifting gears in your shoulders and chest?

Third, name one sensory response (affect tag) you might feel in your body when you make an effective "I," "You," or "Team" statement that constructively reshapes the direction or momentum of a situation. These might include, a sigh of relief, accelerated heart rate due to the excitement of putting yourself forward, faster breath rate due to intensifying frequency of words and thinking.

You have now added emotional intelligence and self-understanding to I, You, and the Team statements. Now that you have some nuts-and-bolts basics for exchanging technical and interpersonal information in engineering communications, you can move into the more subtle aspects of Attending Behaviors that support clarity in delivering content.

CHECK IN

Use this rubric to check in on how well you are developing the I, You, and the Team related microskills.

	1 = Not Attained	3 = Satisfactory	5 = Outstanding
I, You, and, Team Statements	Communication exchanges include criticism and/or blame. Personal responsibility is not articulated for opinions and feelings expressed.	Personal responsibility is expressed. Articulates balanced interpersonal and technical content. Capacity to reorient communication flows toward productive outcomes is not yet exhibited.	Personal responsibility is expressed. Articulates balanced interpersonal and technical content. Reorient communication flows toward productive outcomes.

CHAPTER 9

PAYING ATTENTION WITH ATTENDING BEHAVIORS

Attending behaviors are microskill tabs in our Communication Microskills Model that balance potentially polarized affect processing in interpersonal and technical exchanges. Your skilled Attending Behaviors influence the ideas and emotional responses of others to energize engineering tasks toward successful completion.

You might think communication is just the words. But it is more than that. The words come wrapped up in a nonverbal package. *Attending behaviors* address both the nonverbal package and the words in the package. They allow you to get whole package out of the "I," "You," and "Team" statements made between you and your co-workers.

We start with a dialog that demonstrates how an engineering communication exchange can impact technical content flows when attending behaviors are not used.

Dialog in which you as the listener alter your nonverbal cues to improve the communication exchange.

You are a member of a project team, sitting at your desk and working on your laptop. Your colleague Martha enters the room with a report on her progress for evaluating and costing parts for the upcoming project

Effective Interpersonal and Team Communication Skills for Engineers, by Clifford A. Whitcomb and Leslie E. Whitcomb.
© 2013 by The Institute of Electrical and Electronics Engineers, Inc. Published by 2013 John Wiley & Sons, Inc.

conceptual design review. The project is on a tight schedule. Martha is a highly productive team member and known for engaging in intensive discussions as a way to move the project forward. These discussions are helpful but can run on, so you need to respond to her without getting pulled into a time draining dialog and without alienating an influential teammate.

Martha, "Hi, how's it going?"

Head down, engrossed in your computer screen, you say, "Fine."

Knowing you need the report Martha has, yet feeling pressured to get your own task prepped for the next meeting, you stay engaged with your laptop and ask, "Got your design analysis for the project?" Your voice tone is pleasant and inquisitive, but you have not made eye contact or changed your closed body position—you remain hunched at the desk.

Martha shuffles her feet, clears her throat, and thumbs the papers.

You can ask her to put the papers on your desk, say thank you, and then keep working.

You can wonder why Martha's voice escalates and her frustration boils over in the next team meeting when she tells the team leader, "I did the technical evaluation and costing and took my analysis around but I don't think anybody really integrated it with their own review prep."

You can eel frustrated and upset because you spent some serious time looking at those numbers after Martha left the room, and now you feel accused of not being a team player, and appear unprepared—even though you worked very hard.

Or

You can look up from your laptop, turn in your chair to face Martha, and say, "Thanks, I'll look at these as soon as I can. You do great work, and I need your input to get this project review done right, but I've got to get back to my work right now, prepping these charts is taking over my day."

The eye contact, open body posture, voice tone that matches facial and voice cues and face-to-face acknowledgement of Martha's contribution go a long way to building reserves of goodwill for team meetings and actions. These nonverbal changes acknowledge Martha and give you firm ground to stand on while protecting your time by telling her you have to get back to work. Since Martha is a fellow hard-worker, she will most likely understand the pressures you feel to do a good job, and she will feel that you have acknowledged it in yourself, and for her.

The following example demonstrates a situation for you as the *speaker* to show how attending behaviors can impact the communication exchange.

Dialog in which you as the speaker alter your nonverbal cues to improve the communication exchange.

If you did not communicate your engagement and boundaries clearly to Martha, you may end up in a the team meeting with Martha where she makes a statement such as, "I did the technical evaluation and costing and took my analysis around but I don't think anybody really integrated it with their own review prep."

You attempt to communicate that you did indeed take in her information.

You start by holding your elbows on the table and focusing your eye contact on your team leader.

You say, "Hey, I did read that analysis, and I've got my mine right here—the numbers show some disparity with my design projection charts."

You then lean back and cross your arms, unintentionally glancing at Martha with a blank face as you do so.

Martha's posture becomes stiff, her shoulders hunch, your team leader's face gets that stony look he is famous for when a team meeting goes off track.

You can clear your throat, look down, and go silent.

Or

You can salvage the moment by looking at both Martha and your team leader with a smile, hand your report to Martha across the table and then lean back in a relaxed and open posture.

You can say, "Your numbers were great, Martha—and you brought them to me even before I needed them. Thanks. I am mostly asking a question here, and need some feedback."

Your open posture, smile, inquiring voice tone, and gesture across the table all re-open the door you unintentionally closed by using nonverbal cues that blocked the communication exchange rather than moving it forward.

VERBAL COMMUNICATION DEFINED

In order to begin to use attending behaviors to track your own communications and the impact those communications have on others as fluently as the engineer in our dialog example—you first need to know the difference between verbal and nonverbal communication.

Verbal communication includes the words you use to express technical and interpersonal content. It includes the words you hear when others are expressing themselves to you.

NONVERBAL COMMUNICATION DEFINED

Even though nonverbal communication is hard to measure and define, it is possible to see and track and shape.

Nonverbal communication includes how you posture and compose yourself while you communicate, from your face and your eyes to your whole body, including breathing and muscle tension.

Attending behaviors for both the listener and the speaker involve attending to both verbal and nonverbal communication.

Attending Behaviors for the Listener in Verbal and Nonverbal Modes

You track your teammates nonverbal and verbal information as you listen to

- *their* voice tone,
- *their* body posture,
- *their* visual cues,
- *their* feelings,
- *their* technical content.

Attending Behaviors for the Speaker in Verbal and Nonverbal Modes

You monitor your nonverbal and verbal information as you speak

- *your* voice tone,
- *your* body posture,
- *your* visual cues,
- *your* feelings,
- *your* technical content.

You use *attending behaviors* to track voice tone, body posture, visual cues, feelings, and technical content of your teammates. You take in this feedback to make sure you have both nonverbal and verbal understanding of what people are saying to you.

You use *attending behaviors* to monitor your own voice tone, body posture, visual cues, feelings, and your teammates reception of your technical content so that you know your communication delivery is complete.

NONVERBAL ATTENDING BEHAVIOR FOR THE SPEAKER: SOLER

The foundation of tracking your own nonverbal cues can be found in these behaviors, expressed as SOLER (Egan, 2009).

- *Squarely face them*, let the person you are speaking to know you are fully engaged in the dialog by directly facing them.
- *Open posture*, keep your body posture open, arms across your chest can be perceived as a closed door or a body half turned away can be perceived as a lack of attention.
- *Lean* slightly into the conversation to direct your attention toward fully receiving the person you are listening to.
- *Eye contact*, maintain regular and appropriate eye contact.
- *Relax,* your own presence is enough to move communication forward. Also, when you are relaxed you take in more information.

These basic, nonverbal attending behaviors can be recalled through the use of SOLER.

SOLER

- Squarely face your teammates when they speak,
- Open your body posture while listening,
- Lean towards the person,
- Eye contact maintained,
- Relax your own breath and body posture.

It is important to note Attending Behaviors as practiced through SOLER are presented from a context that follows practices primarily based in Western communication norms. *To be respectful of colleagues and teammates who live/work in other cultural settings, a different approach to attending behaviors may be necessary.*

The steps of SOLER allow you to intentionally receive *all* the information that peers and teammates are sending—including nonverbal, sense/feeling *and* technical/nontechnical cues and words.

But how do you attend to the nonverbal reception happening in others while you are talking? After all, the nonverbal field of transmission of voice tone, body posture, eye contact, and gestural expression includes both the listener and the speaker.

NONVERBAL ATTENDING BEHAVIOR FOR THE LISTENER: RECAP

Just as with the appropriate use of "I" statements, attending to how your verbal statements are being received by others gives you the power to shape the effectiveness of your communication. Attending to the impact of your nonverbal and verbal information is simple and can become an automatic skill.

- *Relax* into your own body posture, notice how the chair feels beneath you or how it feels to be standing to speak—your feet on the floor, your spine supporting your posture.
- *Earmark* your voice tone, as you speak. Do you hear your voice as convincing? As quiet and measured? As assertive?
- *Catch* your impact. How do the people around you who are receiving your communication look to you? Bored and distracted? Engaged? Ready to fire right back at you with a new piece of information?
- *Ask* questions. Check into your listening partner. Simple questions, such as, "How does that sound?" or "Am I making sense to you?" can diffuse resistance to your content and engage the listener very quickly.
- *Pick* up flying cues. What cues is your listener sending that seem random, unrelated, out there or that give you a feeling you are really being heard?

RECAP allows you to intentionally monitor your own presence in the communication, the nonverbal cues that your voice, eyes, senses/feelings and posture convey while you are receiving technical and nontechnical information from peers and teammates.

RECAP

- Relax into your own body posture,
- Earmark your voice tone,
- Catch your impact,
- Ask questions,
- Pick up on flying cues.

It is important to note Attending Behaviors as practiced through RECAP are presented from a context that follows practices primarily based in Western communication norms. *To be respectful of colleagues and teammates who live/work in other cultural settings, a different approach to attending behaviors may be necessary.*

Complete communication reception can be accomplished through SOLER and RECAP practice of attending behaviors. When these skills are integrated into your communication exchanges you are able to practice attending behaviors that move peer-to-peer and team discussions forward with a minimum of time-sapping misunderstandings and a maximum of successful task fulfillment.

When these skills are not integrated, you end up with the consequences of *Missed Cue Attending Behaviors*

Missed Cue Attending Behaviors

Listening

- Turning your head away while someone is speaking to you.
- Looking at your computer, fidgeting with pens.
- Crossing your arms across your chest.
- Sitting with arms on the table and head tucked down.
- Staring fixedly at the speaker with no breaks for head nods or brief verbal encouragers such as, "I hear you," "Yeah," "Okay," "MmmHmm."

Speaking

- Running on for several minutes while not giving a listener time to comment or add a piece of dialog.
- Making brief, declarative statements that are solely fact driven and do not respond to interpersonal dynamics.
- Missing cues of lack of eye contact, body posture, or gestural behaviors given out by the listener to convey that something you have said is missing resistance.
- Delivering information from a tense or rigid body posture, using an angry, critical, or sarcastic tone, making eye contact that is staring and intense or making no eye contact while speaking.

CHECK IN

Can you remember a time when your technical engineering communications were interrupted or rendered inoperable due to missed cue attending behaviors?

How does it feel to consider intentionally attending while listening? Stilted and forced? More engaged?

How does it feel to consider being intentionally attended to while you are speaking? Too visible? Or a nice feeling of being heard? Or somewhere in the middle?

Studies have demonstrated that communication accuracy is supported and increased when attending behaviors are used in many forms of personal and technical communication. So it is important, as when learning to ride a bike, to move through the temporary unbalancing that happens when attending behaviors are first used. Breaking through into communication transmitting and listening fluency is worth the effort.

Try This

Choose a person to have a conversation with about the weather, your favorite sport, pet, food, or a homework assignment.

Practice SOLER while listening

- Squarely face your partner,
- Open your body posture,
- Lean into the conversation,
- Eyes remain on your partner when they are speaking,
- Relax into the conversation, keeping your mind chatter primarily focused on the conversation.

Practice RECAP while speaking

- Relax into your own body posture,
- Earmark your voice tone,
- Catch your impact,
- Ask questions,
- Pick up on flying cues.

Make a checklist for your partner's unique use of these attending behaviors. Have your partner do the same for you.

SOLER and RECAP with Affect

Now try the above but add a little distracting emotion
Talk about something that you or your communication partner may be looking forward to, like a vacation or a work bonus.

Talk about something that reflects your inner values or those of your communication partner, like volunteering for kids in foster care settings or connecting with a grandparent who needs more attention and care lately.

How does it impact your listening skills and selfmonitoring skills when you are observing more intense emotions in your own or a peer's expression of detail and emotional weather around that detail?

What microskills of SOLER and RECAP kept you on track even when cognition and emotional processing shifted balance?

CONTEXTUAL INTEGRATION OF ATTENDING BEHAVIORS

Listening is a foundation of attending behavior. You need to be able to let peers and teammates tell their perspective. When you are the listener you are receiving your teammates' communications to you.

Self-tracking your own output while speaking and listening, including your own nonverbal cues and the nonverbal cues of others, allows you to practice complete communication reception. Complete communication reception is accurate communication reception.

It is more complete and accurate when you attend to both nonverbal and verbal cues because communication is understood by using 87% nonverbal message comprehension (Mehrabian, 2007; Knapp and Hall, 2007; Birdwhistell, 1970) in technical and nontechnical communications. In any given exchange between yourself and peers or yourself and a group, your communication of technical or information-based knowledge is *transmitted and received through a field* of voice tone, body posture, visual cues, and sensory feelings.

Attending behaviors help you navigate this *atmospheric field* of transmission. They do so both when you are transmitting information and receiving information. *Attending behaviors* allow you to accurately understand what others are saying and they allow you to monitor how your words are impacting your listeners.

Accurate communication allows technical information to be related, understood and acted upon for maximum clarity in design and implementation fulfillment.

Attending behavior introduces you to the impact you can have on your communication exchanges by simple sensory tracking, content observation and delivery skills. Next you will learn how to actually start shaping the flow of communications by questions that increase and decrease or change the direction of communication flows in engineering communication.

CHECK IN

Use this rubric to check in on how well you are developing the Attending Behaviors related microskills.

	1 = Not Attained	3 = Satisfactory	5 = Outstanding
Attending Behaviors	Is not receiving or transmitting accurate interpersonal and technical content. Is missing visual and audio cues transmitted by self and others during communication exchanges.	Receives and transmits accurate interpersonal and technical content. Catches visual and audio cues transmitted by self and others during communication exchanges. Clarity in communication flows is not yet present.	Anticipates the impact of verbal and nonverbal missed cues on communication exchanges. Maintains clear and accurate interpersonal and content flow.

CHAPTER 10

 SHAPE YOUR
COMMUNICATIONS USING
OPEN AND CLOSED
QUESTIONS

Open and Closed Questions are microskill tabs in our Communication Microskills Model that give affective and attending behavior skills a structure through which to shape and drive interpersonal and technical communications.

We have established a foundation for interpersonal communication using the microskills of "I" statements and Attending Behaviors. We will now expand into how to pose two specific types of questions, open and closed. We then show how these are used to form attentive communication (including nonverbal and verbal aspects), develop more accurate information, and control the dynamic flow of the dialogue and resulting actions.

Open and Closed Questions Examples

Closed Questions
    Do you get along well with your teammate?
    Have you completed your first design iteration?
    Did you like that meeting yesterday?

Open Questions
    How is it playing out in that struggle with your teammate?
    What are some of the key features of your design iteration that meet the customer's need?

Effective Interpersonal and Team Communication Skills for Engineers, by Clifford A. Whitcomb and Leslie E. Whitcomb.
© 2013 by The Institute of Electrical and Electronics Engineers, Inc. Published by 2013 John Wiley & Sons, Inc.

What did you see that worked in the meeting yesterday?

Engineering Scenario Example

An individual engineer is reporting out to a cost manager.

She might use the following open, attentive questions

"How are we doing with budget predictions for this stage of the design process?"

"How is our feedback on your budget projection working for you?"

Two closed, attentive questions are

"Are we on target for budget limits?"

"Are we giving you our numbers in a way that works for you?"

OPEN AND CLOSED QUESTIONS DEFINED

Open and closed questions extend attentive behavior by asking questions that can control the flow of interpersonal and technical information in a communication exchange.

Open questions ask for general information, and ask you to think and reflect to form an answer. They are open in that they seek answers that are synthetic, inviting someone (or a group) to collaborate with you to put information together into a full and meaningful answer, one that includes the participant's knowledge and feelings.

Closed questions ask for specific information, and can typically be answered in just a word or two. They are closed in that they seek answers that are more analytic, inviting specific points of reference to give you information to address a direct need.

Open and Closed Questions

- control the flow of interpersonal and technical information,
- control the flow of information content in a dynamic and ongoing balance between interpersonal and technical,
- get at two kinds of information, general and specific.

Open Questions

- Ask you to think and reflect to form an answer.

Closed Questions

- Ask for specific information, and can typically be answered in just a word or two.

CHECK IN

Are you more comfortable asking open questions? Or closed questions? What do you prefer to be asked when someone is asking you for information—open or closed questions? To help you reflect on these questions (which, by the way, are both open and closed simultaneously— now figure that one out).

Which are easier to construct in a dialogue? The closed or the open questions? You may have noticed how easy it would be to answer the closed questions with a single word answer. You may also have noticed that in answering the closed questions with a single word answer, your answer would have been less than complete in response to the actual question.

Can you think of an open-ended question? Can you structure a question to ask a partner about the weather, or what they had for lunch that invites them to give you detail, a sense of their perspective and experience? That would be an open question.

Can you structure a question to ask a partner about what they had for lunch that gives you a specific, single word response that allows you to choose, or not to choose to eat lunch where they ate? That would be a closed question. Working with constructing the questions themselves is the best way to learn the skill of both asking and answering open and closed questions.

Ask yourself the following questions:

- Consider a team situation where a team member has not been participating in the team meeting discussion. How do you get your teammate to talk?
- Consider the opposite team situation. How do you get your teammate to slow it down enough to let others participate or how do you get she/he to stick to the point?
- Consider how you feel about forming open and closed questions to deal with those situations.

Try This

1. Think of three open-ended questions you can ask your teammate in charge of design for manufacturing. These questions should elicit answers that give you an idea of a broad timeline, a general design process, and a context overview.
2. Think of three closed questions you can ask your teammate in charge of software development on the project. These questions should

elicit answers that give definitive information, such as deliverable deadlines, budget constraints, or engineering functions.

3. Consider why both sets of data received based on both open and closed questions are necessary to team decision making. What happens when you ask your talkative teammate just a few closed questions? What happens when you ask your monosyllabic teammate several open-ended questions?

4. Consider why this series of questions creates a communication exchange that covers both interpersonal and technical information content.

Open and Closed Questions with Affect

Now think about constructing some of the same questions when there is more pressure in the situation.

For questions 1 and 2 add in that the stakes are now higher because the factory you have contracted with to do manufacture of your design has been severely damaged in a tornado.

For question 3 add in that your talkative teammate shows obvious signs of frustration and anger when you try to limit their responses with closed questions. Their voice gets louder as they interrupt more frequently to keep shaping the flow of your communication to their own comfort level. Their face may get red, their breathing may become more rapid and their posture may become more assertive.

Now for question 3, consider that your monosyllabic teammate may retreat into a wall of silence when you try to draw them out with open questions. They might get very still, their eyes shifting rapidly back and forth, and they might hunch their shoulders and begin answering with shrugs rather than words.

CONTEXTUAL INTEGRATION OF OPEN AND CLOSED QUESTIONS

Open and closed questions operate through asking for general information (open), or through asking for specific information (closed). They are open in inviting answers that include both interpersonal content and technical data. They are closed in inviting answers that offer specific points of reference that are both interpersonal and technical. Open and closed questions are dynamic, allowing you to control or hand control of the communication exchange over to your listeners.

Open questions can shape the flow of information toward including more interpersonal content. Closed questions can shape the flow of information toward including more technical and behavioral actions that follow on the conclusions drawn from using open questions.

Open questions also draw out sensory responses, emotions, role confusion or role responsibility taking. Open questions are very powerful. When you ask an open question, and listen to the answer with SOLER attending behaviors. The response may end up giving you a wealth of the invisible, nonverbal and interpersonal dynamics that are the filter through which all technical information is conveyed.

Closed questions help take this information and funnel the energy generated by open questions into behaviors, operations and actions that move task fulfillment forward. Closed questions are also very powerful when you ask them using SOLER, in a focus that is grounded in both interpersonal and technical understandings of any given situation.

Without the benefits of the information gathered through open-ended questions you are working blindly, similar to trying to breathe air without acknowledging atmospheric conditions on the planet. Without the benefits of the dynamic energy gathered through closed questions you can get mired in interpersonal cul de sacs that inhibit action and task fulfillment, similar to gunning a car engine that is not wired to a chassis and an axel with wheels. Open and closed questions drive discussions because they allow you to shape the flow of affect and cognition as communications unfold.

This is important to note because affect becomes intensified whenever you attempt to shift the dynamic in an exchange. The person who talks too much will not know where to put their energy next. The person who talks too little will not be comfortable putting more open energy into an exchange. To address this, you will need to practice and then practice again the skill of asking open and closed questions, using SOLER and RECAP and choosing appropriate I, You, and the Team, statements, to support your own clarity of response. You will need to bring in the microskill of multimodal attending to juggle technical content, interpersonal feelings, and your own and others' affective processing styles.

CHECK IN

Use this rubric to check in on how well you are developing the Open and Closed Questions related microskills.

	1 = Not Attained	3 = Satisfactory	5 = Outstanding
Open and Closed Questions	Inadequate usage of open or closed questions to elicit necessary information that furthers interpersonal and/or technical communication flows.	Uses open and closed questions to elicit necessary information to further interpersonal and technical communication flows. Ability to shape the flow of communication not yet present.	Uses open and closed questions to elicit necessary interpersonal and technical communication flows that shape productive outcomes.

CHAPTER 11

MOVE INTO MULTIPLE DIMENSIONS WITH MULTIMODAL ATTENDING

Multimodal attending is the microskill tab in our Communication Micro-skills Model that encourages a variety of microskill pairings. Multimodal attending allows all the microskills to work in multiple combinations of pairings and settings.

You may have noticed that the three communication microskills we have just given you juggle several things at once, for instance, attend to someone while considering verbal and nonverbal aspects while also asking open or closed questions. When you add in the technical content at the same time, you are using multiple modes in a more complete engineering communication package. This is the *Multimodal Attending* skill set.

Multimodal Engineering Analogy—Radio Frequency Modulation

Marconi showed that high frequency radio waves could transmit across an ocean, but those high frequency waves had to be mixed, or modulated, with the audio signal to be able to carry the desired information content so far away. To create this radio signal, you need to mix the intended information you desire to get on the airwaves taken from an audio frequency source, such as music or voice, with a second much higher frequency that is meant

---

Effective Interpersonal and Team Communication Skills for Engineers, by Clifford A. Whitcomb and Leslie E. Whitcomb.
© 2013 by The Institute of Electrical and Electronics Engineers, Inc. Published by 2013 John Wiley & Sons, Inc.

for a completely different purpose, to carry energy over long distances. You mix, or *modulate*, the carrier frequency with the audio frequency to create a single new frequency for radio transmission and reception.

Mix the two frequencies together, and you end up with a single frequency that is only slightly different than the original carrier frequency, and it will transmit information over a long distance. At the other end, you can then subtract out the carrier frequency that was used to transmit over that long distance, and you subtract it, what you have left is the original music or words to output at human audio frequency.

In multimodal attending, you are mixing two distinctly different signals together to create one information exchange, communicating technical information along with your affect.

Multimodal attending allows you to accurately transmit and receive technical information while simultaneously tracking interpersonal cues.

Multimodal Attending for an Engineering Project

You are a member of a project team that is developing a robotic vehicle. Numerous feasible alternate product designs can be defined based on variations of many technical considerations. For example, there are many possible designs with different levels of energy storage, power level, payload capacity, and sensors. Each alternate robot design has different operational characteristics, forming a set that a team can use determine which best meets customer requirements. You are in a team meeting discussing these variables.

In the technical mode, you primarily use cognitive processing when considering the technical, cost, risk, and schedule aspects. You use reasoning to maintain the dialog with respect to engineering results. You observe your own thought processes as you reason about your conclusions. You observe the reasoning of your teammates as they dialog with you. These guide you in reaching shared conclusions related technical considerations.

But while you are observing the technical aspects, you are simultaneously observing sensory and affective cues of voice tone and posture in the dialog. These clarify your observations of teammates reception of your information. They also clarify your observations using sensory and affective cues that your teammates use as they state their technical perspectives.

In the interpersonal mode, you primarily attend to your own and others' sensory and affective processes. You might notice that you are feeling frustrated that your good ideas are being resisted, or you see disappointment on the faces of your teammates as their ideas are not taken up—and this gives you pertinent information.

These observations give you information that you can then use so that reactions to technical modes do not block the project forward progress. For example, if you observe that someone on the team keeps insisting on a specific design for the robot, and no one else on the team seems to feel that it will meet the customer need, you can maintain a flow of information using the interpersonal mode. Your observations can be used to state, "I see that you are disappointed that your design isn't being used. I see that the team is frustrated. How can we take care of this?"

MULTIMODAL ATTENDING DEFINED

We are asking you to practice *multimodal attending* to both your own and others' verbal and nonverbal behaviors while you engage in your personal and team technical processes.

Multimodal attending means monitoring yourself and others in an exchange while thinking, speaking, and listening.

Multimodal attending means accurately transmitting and receiving technical and interpersonal content simultaneously.

Multimodal Attending

- considers technical and interpersonal communication modes,
- addresses technical portions and attending behavior,
- considers transmission and reception in both modes simultaneously,
- involves observing—monitoring of self and others.

Because multimodal attending is complex, involving accurate attention to technical, interpersonal, emotional, cognitive, and relationship communications—it happens on a spectrum. The spectrum shifts moment-by-moment in any given engineering communication exchange.

Multimodal Attending Spectrum

Technical Mode
Primary

- uses cognitive processing through critical thinking based on facts,
- validates engineering feasibility.

Secondary

- considers sensory and affective information to bridge with interpersonal modes.

Interpersonal Mode
   Primary

- uses sensory and affective processing of interpersonal exchanges based on individual and team perspectives.

Secondary

- considers cognitive processing to bridge with technical modes.

SENSORY PROCESSING DEFINED

Sensory processing is the physiological transformation of information based on reception and transmission during communication. Electrical impulses from auditory, visual, and kinesthetic stimulation enter your brain while you are speaking, listening, seeing, and gesturing. They are transmitted to areas in the structure of your brain that perform perceptual, associative, and cognitive functions. What you hear, see, and feel (from your body posture and gestures)—is processed using these sensory stimuli. What you say, how you say it, and how others see you posturing and gesturing as you speak is processed in the structure of their brains in the same way. That is shared, sensory processing.

AFFECTIVE PROCESSING DEFINED

Affective processing is the basis of cognitive processing. Developmentally, it is what we learn to do first. We learn to organize sensory stimuli into perceptions, memories, and sensory motor patterns that allow us to monitor ourselves and others. We learn to have emotional associations based on these stimuli. These emotional associations support our basic needs to be attached, be of service or engage in "fight or flight" self-protection.

COGNITIVE PROCESSING DEFINED

Cognitive processing is a specific brain function that happens in the right and left hemispheres of the brain. It utilizes sensory and affective information to translate stimulation and sensing into abstract thinking, language usage, reasoning, and behavioral choices based on thought patterns.

CHECK IN

Consider two aspects of interpersonal multimodal attending that you were not necessarily aware of in your technical content.

Consider two aspects of technical delivery that can be improved by attending to the cues of others as you speak.

How does it feel to consider an attempt at multimodal attending? Natural to what you already do? Completely distracting?

Try This

In your next meeting, observe two other people. Note their unique personal style of listening and speaking. Note their ability to modulate their own inner pacing, tone, visual cues, and content as they speak. Note their ability to respond to each other in a way that acknowledges differences while keeping a balance of interpersonal and technical content and focus.

Multimodal Attending with Affect

Now try the same exercise but this time see if you can note physical cues that are describable as emotions.

- See if you can translate a voice tone into a describable emotion—are you hearing boredom, anger, frustration, neutrality, interest?
- Does the voice tone and your impression of the affect match the technical content or design/operation issues being discussed? Or is it somehow out of context, overdriven or disconnected?
- When you practice this type of multimodal attending you are making it real. You are taking the idea of effective engineering communication and translating that idea into fully dimensional practicable behavior.

You are now ready to actually begin to influence the communication content and outcomes of technical and nontechnical interpersonal exchanges in engineering settings. This will involve learning microskills of encouraging, paraphrasing, summarizing, and reflection of feeling in your exchanges.

CONTEXTUAL INTEGRATION OF MULTIMODAL ATTENDING

Multimodal attending works because *both technical and interpersonal transmission/reception of information occur in a shared field of sensory, affective, and cognitive processing. Unless you actively self-monitor and track others in all three processing modes, you are missing cues and information and losing accuracy in your engineering communications.*

When considering the technical mode of communication, cognitive processing is the primary focus. Technical issues generally require critical thinking related to fundamental science and technology aspects, or other related areas such as cost or schedule, for a project.

When considering the interpersonal mode of communication, sensory and affective processing are often the primary focus. Interpersonal issues generally involve the sensing and expression of emotion, preference, need, and emotional responses to team members and others.

You most likely have no problem communicating technical aspects in a project situation. For example, how much power is needed, what the material properties are, or how costs have been estimated. And we've just told you in previous chapters how to use interpersonal skills to be attentive to the people you are communicating with, and to shape a dialog with the use of open and closed questions.

Multimodal attending means putting this all together. You are then forming the technical aspects into a communication exchange package with an attending behavior wrapper. In this way, you get the technical message addressed while ensuring that the attending behavior considers transmission and reception in both modes, technical and interpersonal, simultaneously. To accomplish this, you need to think about what you want to communicate, and at the same time observe what is going on during the communication exchange.

This is important because observations of multiple modes of content, interpersonal weather, and technical/affective processing of information in an exchange validate or invalidate what is going on to show you whether or not you are moving ahead. Observing also provides guidance on which of the microskills to apply to adapt your exchange to maintain a smoothly flowing conversation and get the results you and your team intend.

This is necessary because if you recall our introduction chapter on self-understanding and emotional intelligence you will remember that these two modes of processing are not physiologically separate. You might remember the three aspects of communication—affect, cognition and behavior, and that all engineering communication exchanges, both technical and nontechnical, occur with 87% non verbal content that originates in our affective processing modes.

Multimodal attending thus becomes an important and powerful skill in effective engineering communication. Multimodal attending gives you the agility to shift from primary cognitive attention to primary interpersonal attention and back again—all within neutral, positive or stressful exchanges of technical and nontechnical information. This allows you to have complete engineering communication exchanges. Ensuring that your engineering messages are received with the clarity that moves designs, projects, and operations forward successfully.

The multimodal attending microskills are often used to dynamically adapt the dialog as you observe resistance and confusion surrounding the technical messages. You can switch to a primary interpersonal mode to gather observations. Attending to your own and others' sensory processes—such as feeling your heart rate get stronger, hearing a voice tone that is more emphatic than you are comfortable with, or seeing your teammates posture close down and their eyes glaze over—gives you pertinent information. These observations demonstrate to you that while you think your technical information is accurate and applicable, the people you are speaking to are not receiving the information accurately and completely. You can repeat the accurate technical details until you are blue in the face—the repetition will not create receptivity.

SECTION III

MAKING IT REAL

Following our DNA analogy, we can say that reading the previous sections and practicing the microskills repeatedly would be the same as aligning the tabs on your own unique communication strand. Now you've got your structure in place and ready to respond interactively with other people to form a communication exchange that is greater than the sum of its parts. The next set of microskills will give you skills to make that synergy happen.

Beginning with this section and continuing through the end of the book we will now leave behind the format of sample dialog, microskills defined and contextual integration. We will invite you into an engineering communication design lab scenario. There will still be some definitions and explaining. But the core of your reading will be in getting an inside view of active engineering communication exchanges and using your own analytical skills in assessing those dialogs for communication self-efficacy.

We will begin by introducing you to the fluent microskills of encouraging, paraphrasing, summarizing as they will be used by Nestor, an engineering team member, to help Lisa—a teammate, deal with another team member whose lack of task completion is impacting her own success on the team.

Effective Interpersonal and Team Communication Skills for Engineers, by Clifford A. Whitcomb and Leslie E. Whitcomb.
© 2013 by The Institute of Electrical and Electronics Engineers, Inc. Published by 2013 John Wiley & Sons, Inc.

CHAPTER 12

DEVELOP FLUENCY WITH ENCOURAGING, PARAPHRASING, AND SUMMARIZING

Encouraging, paraphrasing, and summarizing are microskill tabs in our Communication Microskills Model used in combinations that clarify the synergistic effects of Multimodal Attending. This clarification becomes useful during cycles of problem resolution and conflict negotiation in engineering communication exchanges.

When you read the dialog that follows these definitions, you will see that Nestor knows how to read the cues of his own and others communication exchanges by practicing attending behaviors. You will see that he is also ensuring that he is getting his message heard, that the person he is talking to knows he wants to hear what they are saying, and that he is receiving their communication accurately. He is proficient in a way that can best be described through the following network communications analogy.

Network Communications Analogy

Just as in the case with digital data communication, where a "handshake" is used to confirm that both the transmitter and receiver are aligned in a proper communication "protocol," both people in a conversation need to acknowledge to each other that they are transmitting and receiving to each other on the same channel, using a shared set of rules, to make sure the communication exchange will flow smoothly.

Nestor accomplishes this communication clarity with Lisa through *Encouraging, Paraphrasing, and Summarizing.*

ENCOURAGING DEFINED

You encourage someone in a communication exchange by exhibiting both verbal and nonverbal aspects. Encouraging microskills keep communication exchanges flowing smoothly. It means you are receiving the other person's handshake, know they are there and are interested in sending information back. What are the behaviors of encouraging?

Encouraging

Keeps communication flowing through

Verbal Cues

- repeating key words from the other persons' statements,
- asking for more information,
- saying, "yes," or "go ahead, give me more."

Nonverbal Cues

- nodding your head,
- keeping face-to-face attention wide open,
- smiling.

PARAPHRASING DEFINED

Paraphrasing is essentially you reflecting technical and/or interpersonal content accurately back to the person speaking to you. You reflect back your understanding of the person's meaning, using keywords from their statements to let them know you really listened to their content. This handshake lets them know you are receiving flow and you are aligned enough with their transmission to be able to send it back accurately, with feedback that moves the communication forward.

Paraphrasing

Restate technical content and interpersonal cues—phrases such as

- "I heard you say,"
- "I sensed some frustration,"
- "The design constraint you mentioned."

SUMMARIZING DEFINED

Summarizing is a more complete wrap up of communication flow as a whole, whereas paraphrasing is used in mid-exchange to maintain clarity while flow is happening, summarizing is used to give all involved in the exchange a clear picture of technical and interpersonal content on a specific task or issue. You are giving back larger patterns and dynamics of an exchange rather than the concrete details of an exchange.

Summarizing

Wraps up the gist of an exchange based on encouraging, paraphrasing, and reflection of technical and/or interpersonal content. Uses phrases such as

- "So the picture you gave me is this,"
- "The pieces are coming together but your teammate is holding things up with late design drafts,"
- "You are making technical progress but your team meetings have turf wars happening."

The best way to understand these subtle, yet powerful, microskills is to see them in action between Nestor and Lisa. Read the following scenario to clarify your understanding.

Engineering Project Scenario

Nestor is a member of a five-person project team that is developing a General Autonomous Robotic Device with Expert Networked EffectoRs (GARDENER) to be used to tend plants in a home garden. Along with task assignments, team members have taken roles that help organize team function, such as leader or recorder. His role is that of team task coordinator, making sure your team members complete individual tasks so team function as a whole moves forward.

His teammate, Lisa, asked him to have coffee and talk with her about a teammate who won't do their work. Her task completion depends on his task completion so she is getting behind and getting frustrated.

Nestor is going to use encouraging, paraphrasing, and summarizing microskills to support the discussion.

Lisa

"I'm really frustrated with Joe. My work depends on his work and his work isn't happening. Not only that, he isn't answering my texts or emails."

Nestor (Feels his heart speed up and his muscles go tense as he realizes that a simmering issue between Lisa and Joe is now becoming a head-to-head problem—one that will impact the team and reflect poorly on his own role of team task coordinator. He takes a deep breath and stabilizes his own reaction, noting it for himself so that he can deal with it later. Keeping turned toward Lisa, he nods his head and shifts his shoulders and torso so that his posture is especially open and receptive.)

"I don't like being in that situation either. Have you said anything to him, yet?"

Lisa (getting more animated, leaning forward in her chair, looking more directly at Nestor, then looking down, then looking away and back to him again)

"I ran into Joe in the lab yesterday and told him I couldn't finish my system requirements until I had his customer needs analysis. I've done all the prep I can but I just can't move an inch forward without his information. I told him that. And he flew off the handle and told me that he didn't have time to talk about it right now. He got loud and he got upset."

(Nestor continues nodding his head encouragingly, keeping his posture open and listening carefully so he receives both technical details and interpersonal nonverbal cues. He is simultaneously feeling overwhelmed at the fact that there was an open confrontation between two of his teammates. He keeps breathing slowly, tracking his own posture and facial expression so that for these few moments—his own dismay doesn't interrupt Lisa's process.)

Lisa (gripping her coffee cup hard and tapping her toe really fast on the floor under the table so that it is audible)

"I really want this to come out right for the team. My task is due next week. I don't want to let you guys down. I know Joe can do the customer needs analysis. But right now, we're going nowhere, and if he vents on me again I'm gonna quit the team."

ENCOURAGERS TO USE

Here are some encouraging skills Nestor could use to back Lisa away from the extreme choice of walking away from the team.

Nonverbal

He could include nonverbal gestures and postures while he is listening.

- nod his head,
- keep his posture open,
- practice **SOLER** or **RECAP** skills in his own natural style.

Verbal

Include keywords Lisa used while he was listening.

- "Hmmm, you sound *frustrated*,"
- "Wow, your work depends on his work and his work just is not happening,"
- "He flew off the handle!?"
- "I can tell you really want this to come out right for the team,"
- "Yet I also hear that you still feel you know he can do this so why is he not doing this."

PARAPHRASING SKILLS TO USE

When it is Nestor's turn to speak, he can paraphrase rather than introduce his own feelings and opinions. Adding in his own perspective at this point brings a new complexity or dynamic to the issue. So it is better to keep it simple until the problem is clearly defined. Paraphrasing keeps the focus on the issue and on the person attempting to communicate. He could try something like

"Let me see if I heard you right. You are frustrated because you can't complete your job until Joe does his job and he's not doing it."

SUMMARIZING SKILLS TO USE

Summarizing skills get to the gist of an issue. Nestor could use this skill to reflect to Lisa,

"You want to do a good job for the team and you feel Joe is capable of holding his end up, but you can't get through to him."

ADDING AFFECT TO ENCOURAGERS, PARAPHRASING, AND SUMMARIZING

Nestor can help Lisa diffuse a potential interpersonal, technical, and engineering roadblock by accurately reflecting her technical and interpersonal content while listening to her. But he can begin to actually shape and transform the situation by using affect with his encouragers, paraphrasing and summarizing.

He heard the technical constraints and project deficits being generated by Joe's underperformance. But he can let Lisa know he also heard her feeling—her frustration.

This is important, because it wouldn't be the technical aspect of Joe's underperformance that would cause Lisa to leave the team. It would be **her feelings about** the technical realities.

So why should Nestor care about these feelings? Shouldn't Lisa just get over them?

Nestor needs to care about these feelings because.

- Lisa and Joe are his teammates. *Their performance impacts Nestor's performance.* And even if he works perfectly, keeps his own feelings clear and cooperates at every opportunity. Not everyone else operates that way.
- He is stuck integrating his engineering excellence with people who may underperform, undervalue, or disrupt his technical contributions or allow their own messy emotions to impact his performance.
- He is stuck sharing his engineering excellence with people who may need his reflection and feedback to balance their own affective processing, cognitive clarity, and behavioral choices.
- So learning to add affect to his use of encouraging, paraphrasing, and summarizing—and learning to track the affect of others as he does so when frustrations and differences arise, is a crucial microskill he can use to shape the flow of effective engineering communications for himself and his teammates.

This is important because when people are reflected in their experience of feeling about a situation, they tend to diffuse escalation and stay out of polarizing positions. Our next section—*Reflection of Feeling*—helps you understand why.

Try This

First, think about those team meetings where an over directive peer listens to you ask for airtime to speak. They nod their head vigorously and move right back into dominating team conversations, decisions, and directions.

Second, think about those team meetings where a frustrated teammate vents about how their ideas and input aren't getting enough airtime, in either implied or openly critical interpersonal dialog. The venting keeps getting interjected into technical choices and this sidetracks solutions and forward progress.

Third, consider what derails work and creates tension in these situations. It is the synergy of technical requirements and human response choices for dealing with those requirements. This synergy can move toward stagnation or escalate into conflict especially when engineering processes are not working, working too slow, or working with outcomes that are off target. Then affective processing—dealing with the emotions that come up when you are under pressure—can disconnect or overdrive accurate or reasonable cognitive choices. This will impact technical engineering outcomes.

Fourth, think about how you diffuse tension in a team or home setting. Do you joke? Change the subject? Focus harder on technical details until things cool down? Talk to a friend? Take a run or cook a meal?

Fifth, consider how do you want to be dealt with when your affective processing is clouding your cognitive clarity and behavioral choices, or when you need extra explanation from a professor or supervisor to get it right, or when you are worried about task completion and other commitments are getting in the way. How do you want others to respond to your technical and interpersonal needs at those times?

CHAPTER 13

CLOSE THE LOOP WITH
REFLECTION OF FEELING

Reflection of feeling is the microskill tab in our Communication Microskills Model that allows you to shape a communication exchange even as interpersonal and technical content is cycled through multiple levels of intensive information reception and information feedback.

Ignoring and sidelining the emotions of yourself or others that are generated in creative engineering processes doesn't make them go away. Joe would have liked to sideline Lisa's inconvenient reactions to his lack of input on team tasks. Lisa would have liked to sideline Joe's response to her. But in doing so, they both would have missed crucial information necessary to enable them to work together in accomplishing a shared engineering task.

When you are stuck with a project task that presents seemingly unresolvable technical issues or when you are stuck with a teammate who simply will not perform or whose performance disrupts your own. You can use *Reflection of Feeling* to help yourself and others move past obstacles in task performance and communication.

Effective Interpersonal and Team Communication Skills for Engineers, by Clifford A. Whitcomb and Leslie E. Whitcomb.
© 2013 by The Institute of Electrical and Electronics Engineers, Inc. Published by 2013 John Wiley & Sons, Inc.

REFLECTION OF FEELING IN THE DIALOG BETWEEN LISA AND NESTOR

The keywords used in the dialog between Lisa and Nestor, the words that reflected her feelings about technical needs were

"frustrated,"

"flew off the handle,"

"the good of the team,"

"he has the capability."

These words and phrases represented emotions that are currently in only a semi-functional balance with cognition and behavior. The emotions are becoming a primary focus—pushing the technical considerations into the background by annoying one team member, and creating some engineering flows to be compromised.

If allowed to simmer or continue to be reflected in unconstructive exchanges between Joe and Lisa, these emotions will create an interpersonal atmosphere about technical processes that will negatively shape technical cognition and task-oriented behavior.

These feelings need to be heard, seen, and reflected constructively to diffuse their charge enough that they can then become clear affect responses that drive team tasks.

WHAT HAPPENS WHEN FEELINGS ARE NOT REFLECTED

If these feelings are not reflected,

Lisa may alienate Joe so far that his contribution to team tasks will be nullified. He may be reassigned to another task when his skill level at the task he is holding back on is the highest in the group. This would be a loss of good engineering skills happening in the right time and in the right role for the team as a whole.

Lisa may leave the team if Joe vents on her again. Lisa's considerable engineering skills will no longer serve the team.

Engineering tasks will be sidelined and delayed to the detriment of deadlines and final outcomes for all team members.

So how can Lisa, Nestor, and Joe track their own emotions and accurately reflect the emotions of their teammates in constructive feedback that gets engineering tasks back on the road to completion?

TRACKING REFLECTION OF FEELING

They can begin by learning to track the 87% nonverbal aspect of all technical and interpersonal communication. This will give them helpful clues in reflection of invisible yet palpable feelings that block the accomplishment of engineering goals. The nonverbal aspects in the dialog above showed Nestor that

- Lisa doesn't like to be frustrated (she was straight and tense in her chair, she gripped her coffee cup, her eyes darted from side to side, expressing fear and anxiety).
- Lisa doesn't want to be dumped on for simply trying to get a team member to cooperate (as she related Joe venting on her—her voice got high, rapid, and tight. Her face grimaced, her eyes grew wider with concern, her shoulders hunched).
- Lisa doesn't want to disappoint her team or have her grade or job at risk because someone else won't perform (Lisa's shoulders slumped, her voice got quiet and sad, her speech got slower and more monotone, her cognition slowed way down and her choices of words became more simplistic and less analytical and technical).
- Joe withdrawing from work is a nonverbal behavior that is expressing some distress on his part with team function. Lisa walking away from the team is a nonverbal behavior that impacts the team.

BENEFITS OF TRACKING REFLECTION OF FEELING

Nestor listened fully to Lisa and observed her feeling accurately enough to reflect it back to her in a nonjudgemental, constructive way. His reflection of her experience matched her experience of the situation. By responding to Lisa this way, Nestor gave a full-on demonstration of the fact that,

"Group emotional intelligence isn't a matter of catching a necessary evil—catching emotions as they bubble up and promptly suppressing them. Far from it. It's about bringing emotions deliberately to the surface and understanding how they affect the team's work. It's also about behaving in ways that build relationships both inside and outside the team and strengthen the team's ability to face challenges.

Emotional Intelligence means exploring, embracing, and ultimately relying on emotion in work that is, at the end of the day, deeply human" (Druskat and Wolff, 2008).

Nestor used his own emotional intelligence and modeled this skill for Lisa through encouraging, paraphrasing, and summarizing microskills to which he added reflection of feeling. He made it possible for Lisa to calm down enough to not walk away from the team or initiate another unconstructive go-around with Joe.

FEEDBACK LOOPS THAT BALANCE FEELING AND THINKING: ACCURATE REFLECTION OF FEELING

For those few moments with Lisa, Nestor was a *sensei* of emotion—a master of feeling and reflection of feeling in ways that furthered cognitive clarity and constructive behavioral choices to further team task completion.

He was flooded with that sinking feeling you get when you watch others mistakes start to create a train wreck in your own task completion. As team task coordinator, this lack of cooperation between Joe and Lisa would end up directly reflecting on Nestor's performance.

Yet, even given this added situational stress, he stayed calm and focused on Lisa, stabilizing his own feelings so that he could be present to her. He was able to read his own breath rate, posture, muscle tone, voice tone, and facial responses—to tag and understand and intentionally modulate his own emotions. Even and especially when those emotions were unpleasant and risky.

This allowed him to become present enough to read the cues of Lisa's emotions through noting her breath rate, posture, muscle tone, voice tone, and facial response changes throughout technical and interpersonal communications. He practiced intentional management of internal and external reflection of feeling.

Internal Reflection of Feeling

Because Nestor was capable of practicing internal reflection of feeling for his own internal responses. He would have been capable of observing and mediating the internal reflection of feeling happening for Lisa in the lab with Joe. He may have observed whether or not Lisa

- tagged her body responses to see if her shoulders were hunched, or if her stance was openly assertive, or if her voice was high and intense?
- tracked and then modulated her own internal affect, her frustration with Joe and fear of disappointing her team?
- or did she allow her internal feelings to overdrive her discussion with Joe?

Nestor, in his discussion in the coffee shop with Lisa, modeled a way she could have edited her affect in her dust up with Joe—he modeled a way that she could feel and understand her own frustration and fear, then modulate her affect to inform the discussion rather than polarizing and shutting it down. Following this modeling, she could have

- been able to see clearly through her own affect, and note changes in body posture, voice tone, facial expression and attention that Joe exhibited well before he "flew off the handle,"
- been able to either handle her delivery differently or choose to deal with Joe in a group situation with the back up of teammates rather than on her own.

This is important, because with each overdrive or disconnect of affect in work and interpersonal situations, people have been studied and found to be at risk for polarizing their positions and entrenching themselves in group roles and interpersonal responses—even when those roles and responses are not self protective or constructive for the group as a whole.

External reflection of feeling is equally important in transforming this polarization into team cooperation instead of team derailing.

External Reflection of Feeling

Reflecting the feelings of others is based on a blueprint that can be adapted to many different work and interpersonal situations. Like a protocol for dealing with a fire alarm in a public building, reflection of feeling is a practice that can be reduced to simple steps that can be followed despite situational stress and risk. These steps keep communications effective whether the alarm is just for a fire drill or used to alert a group to a real fire in process. The steps are

1. *Self-Stabilize*: by tagging your own breath rate, body posture, voice tone, facial expression and muscle tone. Trace your own frustration, worry, anger, surprise, desire to help or take charge.

2. *Shift Your Focus*: from your internal reflection of feeling to listening to communication content while observing posture, gestures, voice tone and facial expressions of the person speaking. Listen carefully to the speaker and observe their emotions.

3. *Gain an Accurate Picture*: use reflection of feeling to gain an accurate picture and diffuse disconnected or overdriven affect in the situation that could cloud this picture. You paraphrase and summarize feelings to shift the discussion toward the kind of self-understanding that allows your communication partner to find behavioral choices and strategies

that diffuse escalation and prevent affect from overdriving or disconnecting future discussions.

To illustrate these steps, we will take you back to Lisa and Nestor. These are the steps Nestor followed to remain attentive to Lisa even while he himself was experiencing stress.

Step one is to **self-stabilize**.

To do this, begin by tagging your own breath rate, body posture, voice tone, facial expression, and muscle tone. Trace your own frustration, worry, anger, surprise, desire to help or take charge. Use this self-understanding and emotional intelligence to take a breath, and give yourself permission to protect your own interests while also being of service to others.

For instance, Nestor may have felt any of these emotions as he listened to Lisa,

"Oh, no. Here we go again, another team, another conflict, how is this going to impact my performance. . . . "

"Lisa is a great teammate, I really want to back her up and make sure she gets to do her job."

"I am the team coordinator and I know how to coordinate flows of information but I don't even know where to start coordinating interpersonal bad weather."

To accurately reflect Lisa's experience in the midst of this, he first stabilized his own affect. He modulated his own reactions and adjusted his external listening cues to reflect this stability and attention.

Step two is to shift your primary focus from your internal reflection of feeling to listening to communication content while observing posture, gestures, voice tone, and facial expressions of the person speaking. **Listen carefully to the speaker and observe their emotions**.

In this case, Nestor observed Lisa. She was clenching her coffee cup really tightly while she spoke. Her shoulders are hunched. Her speech was rapid, staccato, and intense. Nestor had to assess these cues carefully to decide if this looked like an annoyed reaction to a typical team frustration or if it looked like something that was really getting to Lisa enough at a very personal level that it would actually cause her to walk away from the team?

Step three is using reflection of feeling to **gain an accurate picture and diffuse disconnected or overdriven affect** in the situation. Nestor chose similar phrases as those below to gain an accurate picture and shape the discussion toward self-understanding on Lisa's part

- "I hear that you are frustrated, but you also look really tense and upset."
- "You sound like you are more than frustrated. Are you scared of Joe or just really mad at him?"

- "I hear you say you are frustrated and that Joe has a lot to contribute. But you sound and look like you are at the point where you maybe don't want to deal with him at all anymore. I don't think I've ever seen you this upset."

Each of these reflections of Lisa's feelings gives her a chance to *reflect on her own experience of the situation.*

That is the magic of reflection of feeling. Your internal stability + your external accuracy in reflecting feeling = clarifying reflection.

This allows teammates to minimize affect, and redirect and reconnect cognition and behavior so that a problem solution rather than a problem escalation can happen.

Nestor reflected enough feeling that Lisa felt seen and heard. Leaving the door open for next steps in actually resolving the issue, not just stabilizing it. This will involve a six-step problem solving cycle that we will introduce in Chapter 14.

CHAPTER 14

THE SIX-STEP CYCLE

The Six-Step Cycle is the microskill tab in our Communication Microskills Model that combines and recombines your own microskills usage with that of others during resolution focused, interpersonal and technical engineering communication exchanges.

Lisa and Nestor worked hard in their dialog to use all the microskills learned so far in this book. They demonstrated a working knowledge of

- practicing self-awareness,
- using emotional intelligence,
- using appropriate I, You, and Team statements,
- asking open and closed questions,
- using attending behaviors,
- practicing multimodal attention,
- using encouraging, paraphrasing, summarizing,
- using reflection of feeling.

In the current situation facing their team, they need to know how to not only shape the flow of engineering communications. But to learn to shape those flows to activate themselves and a colleague, individually and for their

Effective Interpersonal and Team Communication Skills for Engineers, by Clifford A. Whitcomb and Leslie E. Whitcomb.
© 2013 by The Institute of Electrical and Electronics Engineers, Inc. Published by 2013 John Wiley & Sons, Inc.

project as a whole, as they move through cycles of creativity, conflict, task performance, and goal completion.

Being able to shape and drive engineering communication flows in such a holistic way will require them to use the *Six-Step Cycle for Technical and Interpersonal Communications*.

The *Six-Step Cycle for Technical and Interpersonal Communications* is a use of microskills that is multilayered and can be used across diverse engineering contexts. It can be followed repeatedly over the lifetime of a project cycle or division development. It can be used during creative teamwork, in moments of peer-to-peer learning and professional development and in team conflict amid task performance.

This six-step cycle balances nonverbal emotional dynamics, cognitive clarity, and behavioral choices. Mastering the cycle gives the user a communication fluency that makes them adaptive, flexible, and intentional engineering professionals on all levels of engineering communication.

Our evolution of this robust engineering communication structure can be highly effective over a broad diversity of personal and management situations for one very special reason. This reason is that the emotional intelligence of engineers—their capacity to accurately focus on *identifying* an opportunity or a challenge, as well as their capacity to move very quickly from cognition to behavior—creates a powerful hybrid of interpersonal insight and engineering technical functionality.

To maximize this strength common to many engineers, we evolved the six-step cycle to reflect both traditional engineering approaches to problem solving. We evolved the cycle to also reflect the growing needs of the profession for a better balance of interpersonal communication dynamics with technical communication dynamics.

A typical model for solving engineering problems involves three simple steps, repeated iteratively until a feasible solution is achieved.

1. define the problem,
2. define the goals,
3. generate alternate solutions and select an effective engineering feasible approach.

A review of both this more typical engineering model and our own, shared below, shows you a balance of *affective* processing—which is necessary for effective listening and reflection in any situation—and *cognitive* processing—which is particularly necessary in an engineering model. Combining the two works to turn this into *behavioral* actions—which are necessary to actually solve societal needs through effective engineering. You do this through inclusion of a communication practice that is much more than the sum of its working components.

SIX-STEP CYCLE FOR INTERPERSONAL AND TECHNICAL COMMUNICATIONS

1. identify context,
2. define the problem,
3. define the goals,
4. generate alternate solutions,
5. take action,
6. iterate.

The value in this approach is that it balances technical and interpersonal communication and it can be iterated—repeated as needed over the life cycle of a design development or project execution.

Now let's take you from what you know to what you may not know. That is, what you will do in engineering contexts when you are trying to establish intentionality and shape the flow of communication, as you head toward driving your communications rather than *being driven by them during problem solving moments.*

Six-Step Cycle

1. *Identify Context*:   Establish rapport and understand the area of the space, face, and place spectrum in which you are interacting.
2. *Define the Problem*:   What are concerns, issues—define what to talk about—primary technical, a balance of technical and interpersonal, or primary interpersonal. Include attention to interpersonal dynamics that may be obscuring technical solutions.
3. *Define the Goals*:   What do you want to happen from the communication.
4. *Generate Alternates*:   Explore alternates of more effective communication and behavior to create intentionality in the situation.
5. *Take Action*:   Choose an action and follow through.
6. *Iterate*:   Repeat as necessary to achieve holistic action follow through.

Table 14.1 summarizes the six steps to include related communication microskills and predicted results for each stage.

TABLE 14.1   The Six-Step Cycle

Step	Description	Related Microskills	Predicted Results
Identify context	Acknowledge interpersonal and technical relationships and understand areas of space, face, and place spectrum in which you are interacting	Self-understanding Attending behaviors	All team members are interpersonally present and engaged and ready to discuss technical issues
Define the problem	What are concerns, issues—define what to talk about—primary technical, a balance of technical and interpersonal, or primary interpersonal	I, You, and the Team Open and closed questions Attending behaviors Encouraging Paraphrasing Summarizing Reflection of feeling Microskills of affect	The problem gets defined without being overdriven by interpersonal blame or conflict. A clear picture of technical needs is developed
Define the goals	What do you want to happen from the communication	Self-understanding I, You, and the Team Attending behaviors Reflection of feeling Clear affect processing	Interpersonal needs are clearly processed and balanced. Steps for technical task completion are evaluated
Generate solutions	Explore alternates for more effective communication and optimal technical choices to create intentionality in situation	Open and closed questions I, You, and the Team Clear affect processing	Choose technical solutions that include balanced interpersonal dynamics
Take action	Choose an action and follow through. Repeat the steps when and as needed to support this action	Encouraging Paraphrasing Summarizing Reflection of feeling Repeat other microskills, as needed	Interpersonal weather stays clear. Choices for technical task completion get done
Iterate	Repeat as necessary to achieve holistic action follow through		Assess results as you iterate until you converge

ENGINEERING PROJECT SCENARIO REVISITED

Seeing this structure in action is the best way to understand its value. Let's get back to Nestor, Lisa, and Joe. Here is how to understand their situation in our Six-Step Cycle. Let's start back at the discussion between Lisa and Nestor in the café.

Step One: Identify Context

On the Space, Face, and Place spectrum this interchange is primarily interpersonal—you are meeting in a situation that is not a lab, design team, team meeting, or project presentation. Nestor and Lisa have established some level of trust and rapport because they are teammates. They have already undergone some team development period, so they've spent a fair amount of time communicating and know each others' styles and patterns.

Step Two: Define the Problem

Lisa did this when she said, "I'm really frustrated with Joe. My work depends on his, and his work isn't happening. Not only that, he isn't answering my texts or emails."

Note: A common problem in the five stage structure is that the person defining the problem *gets stuck in the concrete details* of problem definition. An important skill to master while intentionally driving the flow is in getting the person you are dealing with to *move on* to steps 3, 4, and 5.

In other words . . .

> If you can't tell me what you'd like to be happening . . . You don't have a problem yet. You're just complaining. A problem only exists if there is a difference between what is actually happening and what you desire to be happening.
>
> —Kenneth Blanchard

The content of Lisa and Nestor's dialog from the chapter on encouraging, paraphrasing and summarizing, stops at problem definition.

Step Three: Define the Goals

Nestor shaped the flow of communication away from problem definition and toward action in an exchange the two had after breaking to get more coffee.

Here's how the dialog exchange played out.

Recall that Lisa has just made her statement that Joe is giving her his cost analysis and won't answer her calls and texts. And that she is stumped for a solution after his blow out.

Nestor

"Lisa, I hear what you are saying. But what do you want to do about it?"

Lisa (first responds to this by simply restating the problem),

"Well, Joe is just not doing his work so I can't do mine."

Nestor, (shaping flow toward goal setting),

"Okay that defines the problem. Now how do we find a solution?"

Lisa (pulling flow back toward problem definition but giving a cue for goal defining)

"Well, I don't want to have to report to the professor in order to get Joe to do more of the work with me."

Nestor, (again shaping the flow toward generating goal definition),

"Yeah, I get that. That is what you don't want as a solution. But what would work better for you than that?"

Lisa, (a potential goal is generated)

"Well, I don't know. Maybe it would be better to deal with Joe in a team meeting. It's really a team issue as well as a problem between me and Joe isn't it? After all, if I don't get my work done because of Joe, it won't look good, and the whole team fails."

Step Four: Generate Alternates

Now a goal has been potentially defined. The goal is to address Joe's impact on team success without Lisa taking the heat. This goal needs to be accomplished in such a way that it does not force Lisa to drag the professor into this and make herself look like she was the one who couldn't deal well with team issues.

Generating a flexible alternate to meet the challenges of this goal comes next and demonstrates Stage Four.

Nestor, (shaping the flow toward alternates that preview outcomes),

"So how do you feel about bringing this up in a team meeting? If Joe gave you heat when you brought it up before, do you think you will have it any easier within a group with him?"

Lisa, (responding to this shaping with confusion and overdriven affect),

"Oh wow, I didn't think of that . . . You know, maybe I just don't fit this team and I'd better get reassigned before the project gets too far on."

Nestor, (reflecting feeling to diffuse it and reshaping the communication toward generating alternates),

"Yeah, I'd want out, too, if it was me. I can see he really blew it and you don't want to have him vent like that again. How about if it's an agenda item and I bring it up as coordinator? I think Sunil will get with us on this in a way that doesn't embarrass Joe. She's a great team leader. I've seen her handle this kind of stuff before and not crash."

Lisa, (responding to reshaping),

"Yup, she did a great job last year with that team that almost fell apart. I can work with that."

Step Five: Take Action

Through modeling all the microskills you have just shaped engineering communication flows. Lisa is not going to quit the team and take her needed technical skills with her. Joe's participation will be addressed in an approach that doesn't include blame and labeling, and the team will benefit by his technical contribution.

The crucial step of Take Action, and Iterate, comes next.

Nestor,

"Okay, if you put the item on the agenda as a general discussion, I'll follow up with Sunil and let her know we might have task completion failure if we don't cover the agenda item."

Lisa,

"I'll go enter the agenda item right after I leave here. I don't want to take a chance I'll get busy and forget."

Nestor,

"Okay, I'm putting it in my phone now to tag Sunil when I see her in lab tomorrow."

Lisa,

"Thanks, Nestor. This is a little over and above your task coordinator job and I really appreciate it."

Nestor,

"No problem, I've been there and I get it."

Step Six: Iterate

Go back over what you have done so far, and see if there are any areas where you need to work out additional considerations. Lisa and Nestor fulfill their tasks and simply continue to bear in mind the need to both bring this issue up at the meeting and keep the meeting on track toward creative resolution rather than creative flame out.

Nestor and Lisa have now demonstrated the best of interpersonal communication effectiveness; the rapport, insight, team loyalty creation, and professional ethics that are possible when two engineers communicate fully and effectively. Lisa, a teammate with necessary skills that uphold technical task completion is still part of the team. Joe has not been verbally trashed and will be given the respect of addressing his issues fairly. Additionally, an interpersonal issue that could stop task completion is on its way to being removed as a roadblock.

Two members of the team have now fulfilled a key proficiency highlighted by Druskat and Wolff in an article in the Harvard Business Review based on their research about building the emotional intelligence of teams.

> "To be most effective, the team needs to create emotionally intelligent norms—the attitudes and behaviors that eventually become habit—that support behaviors for building trust, group identity, and group efficacy. The outcome is complete engagement in tasks." (Druskat and Wolff 2008)

We can hear you thinking that humans aren't perfect and life isn't that easy. You are so right. The confrontation and conflict that ensue when the next team meeting happens for Lisa, Joe, Nestor, and Sunil is not pretty. Instead of a conflict that causes failure, however, it becomes *confrontation and conflict that generate productive outcomes while building professional community*. This is very different than confrontations and conflicts that block or compromise engineering tasks and project success.

CHECK IN

Use this rubric to check in on how well you are developing the six-step cycle and its related microskills.

	1 = Not Attained	3 = Satisfactory	5 = Outstanding
Six-Step Cycle	Does not create workable problem solving context. Does not define goals to adequately address problem. Does not explore alternates. Does not choose an action for follow through.	Establishes rapport and understanding. Defines concerns and issues while accepting personal responsibility. Defines desired outcomes. Explores alternates. Chooses actions and follows through.	Establishes rapport and understanding that actively engages others and invites collaboration. Reflects the big picture of concerns and issues. Elicits and defines desired outcomes. Creates intentionality in exploring alternates. Obtains consensus on potential actions that empower everyone to follow through.

SECTION IV

TAKING THE LEAD

> Leadership is, as much as anything, an emotional adventure. If you want to be a powerful leader, you have to become familiar with the sweat-inducing, anxiety-producing, adrenaline-generating emotions of being lost while people are following you. Because that is, as often as not, the emotion of leadership.
>
> —Bregman (2012)

Taking the lead means influencing outcomes and inspiring others to act toward productive outcomes regardless of interpersonal and technical obstacles. Through taking the lead, you are becoming an intentional engineer.

This is all fine except for the fact that, in an analogy according to Newton's Third Law, for every action there is an equal and opposite reaction. In the same vein, every productive human action results in a human reaction, some of which are productive, but some of which are potentially unproductive. This means that for the life cycle of your career you will need skills to handle confrontation and conflicts that arise from reactions to the actions you choose while taking the lead in engineering contexts.

To do this, you need to be able to *handle confrontation*, *negotiate conflict*, and *influence others*. You will develop your own personal style of intentional engineering communication as you learn these skills. You will then become a more complete and an effective leader in the engineering profession and a highly valued member of your organization.

Effective Interpersonal and Team Communication Skills for Engineers, by Clifford A. Whitcomb and Leslie E. Whitcomb.

CHAPTER 15

WORKING WITH CONFRONTATION AND CONFLICT NEGOTIATION

Confrontation and conflict negotiation are microskill tabs in our Communication Microskills Model that provide you with codes to master some of the most challenging aspects of effective communication.

Confrontation and conflict have been studied and found present in almost all workplace and teaming settings—technical and nontechnical. That means this is a force of human interaction that is going to be present in a work setting regardless of whether you want it to or not. So you may as well know how to use this force for technical development rather than technical obliteration. In this chapter, we show you how to achieve confrontation and conflict negotiation skills that lead to technical development progress and task completion.

CONFRONTATION

Merriam Webster defines *confrontation* as "The act of confronting, the state of being confronted, as a face-to-face meeting, the clashing of forces or ideas; conflict. Additionally, Merriam Webster defines *conflict* as, "Fight, battle, war. Competitive or opposing action of incompatibles, antagonistic state or action (as of divergent ideas, interests or persons); mental struggle resulting from incompatible or opposing needs, drives, wishes, or external or internal demands."

Effective Interpersonal and Team Communication Skills for Engineers, by Clifford A. Whitcomb and Leslie E. Whitcomb.
© 2013 by The Institute of Electrical and Electronics Engineers, Inc. Published by 2013 John Wiley & Sons, Inc.

These elements will unfold in the next team meeting that Nestor, Joe, and Lisa will attend. Joe and Lisa's original face-to-face meeting failed to produce resolution. Engineering processes are now interrupted. Work is not being accomplished. There was a clashing of ideas. Lisa wanted to talk about getting Joe to complete his work. Joe did not want to talk about it. The next meeting will therefore be tough. But it does not have to become a roadblock to constructive technical and interpersonal communication or task fulfillment.

Some of the team members in the next meeting are going to use constructive confrontation and conflict negotiation skills as a lever to shift the roadblock of conflict standing in the way of their task completion. They are going to do this by practicing confrontation and conflict negotiation that highlight and then rebalance discrepancies. This will work because . . .

- The core of unproductive confrontation is that it polarizes discrepancies between perspectives of those involved in any given situation.
- The core of unproductive conflict is that it entrenches discrepancies among differing perspectives, needs, and resources.
- Productive confrontation realigns discrepancies, turning them into complimentary opportunities for action.
- Productive conflict releases the tension held in discrepancies, liberating energy for forward momentum.

When you know how to highlight discrepancies and reintegrate them, you know how to use the dynamics of confrontation and conflict negotiation to lever a roadblock to *enable flow* instead of *blocking flow.* This involves practicing microskills to help move a situation to a successful outcome. When you do this you are not going against yourself or others. You are going *with* yourself and others to resolve difficulties and/or maximize creative technical engineering potentials.

The concepts of carefully levering blockages rather than exploding them, and of going *with* rather than *against* yourself and others, are the heart of our approach to microskills usage in confrontation and conflict negotiation. Using these microskills in confrontation during engineering communication exchanges means addressing

- discrepancies between *goals* and *actions,*
- discrepancies between *intended* meaning and *expressed* meaning,
- discrepancies between *technical* needs and *available* resources.

You do this by following four steps that we would advise you to hold on to as an anchor in a stormy sea whenever your intentional engineering actions meet unproductive reactions.

Confrontation Guide

Step One

Identify Discrepancies

Internally: Self stabilize by tagging your own sensory cues and by observing the clarity of your own cognition and behavior. Use breathing, posture, thought focus, and muscle relaxation to keep yourself stable in our own body. *When you do this, you are preparing yourself to catch your own discrepancies in approaches to a conflict before those discrepancies are communicated to others and create roadblocks to solutions.*

Externally: Use SOLER, RECAP, Attending Behaviors, and Multimodal Attending to catch and redirect the interpersonal feelings or technical inaccuracies others send your way in creative or confrontational situations. *When you do this, you are identifying discrepancies in others' approaches to a problem.*

Step Two

Point Out Discrepancies

Internally: Be aware, be very aware, of your own internal emotions, cognitions and behavior choices. *Look for discrepancies between what you think you are saying and what others are hearing you say.*

Externally: Use the Six-Step Cycle to highlight discrepancies to colleagues and peers in *nonjudgemental and constructive ways. Look for discrepancies in goals and behavior, discrepancies in interpersonal choices and technical outcomes, and discrepancies in technical and interpersonal resources available for the goals identified.*

Step Three

Work Toward Rebalancing Discrepancies

Internally and Externally—Choose and perform tasks that modify and reintegrate discrepancies. *Reflect and support teammates as needed.*

Step Four

Evaluate Outcomes and Iterate

If at first you do not succeed, try, try again

If need be, new discrepancies are identified and put through the same sequence. It is a cycle—not a linear process. *A continuing iterative process of evaluating results in interpersonal, technical and project outcomes based on agreed upon actions and consistently respectful interpersonal responses.*

An example of confrontation that did not work this way was the confrontation Lisa had in the lab with Joe. To refresh your memory about that dialog, we reprint it here for you. In this reprint, we have included steps from our *Confrontation Guide* that occur as the scenario unfolds. These steps are in italics.

Engineering Project Confrontation Scenario I

Nestor's teammate, Lisa, asked him to have coffee and talk with her about a teammate who will not do their work. Her task completion depends on his task completion so she is getting behind and getting frustrated. Nestor is the team coordinator for task collaboration and coordination.

Lisa

"I'm really frustrated with Joe. My work depends on his work and his work isn't happening. Not only that, he isn't answering my texts or emails."

Step One: Lisa has identified an external discrepancy—the gap between Joe's work commitment and the work he is doing.

Nestor, (nods his head and shifts his shoulders and torso so he is facing Lisa more directly, uncrossing his arms so his listening posture is fully open)

"I don't like being in that situation either. Have you said anything to him, yet?"

Lisa (getting more animated, leaning forward in her chair, looking more directly at Nestor, then looking down, then looking away and back to Nestor again)

Step One: Lisa should identify her internal discrepancy here. Her nonverbal cues are broadcasting distress. She may perceive herself as only discussing technical details of task completion. But she is transmitting these in a nonverbal field of interpersonal distress that may obscure her technical message. It is okay for her to have the feelings, she just needs to identify the discrepancy of how she matches her interpersonal delivery with her technical content.

"I ran into Joe in the lab yesterday and told him I couldn't finish my system requirements until I had his customer needs analysis. I've done all the prep I can but I just can't move an inch forward without his information. I told him that. And he flew off the handle and told me that he didn't have time to talk about it right now. He got loud and he got upset."

(Nestor continues nodding his head encouragingly, keeping his posture open and listening carefully so he receives both technical details and interpersonal nonverbal cues)

Step Two: Point Out Discrepancies

Internal: Nestor has aligned his internal discrepancies. Even though he may be feeling overwhelmed by a problem that he, as coordinator, has to deal with. He tagged his sensory affective cues—felt his breath get tight and short, felt his heart rate go up a bit, observed his thoughts racing to find a quick solution. He then selfstabilized internally by slowing his breath and focusing on the potential of a solution, remembering that both Joe and Lisa are hardworking, high performing teammates who care a lot about the project. This helped his interpersonal weather to not overdrive his perceptions and subsequent reflection of discrepancies in his teammate.

Lisa (gripping her coffee cup hard and tapping her toe really fast on the floor under the table so that it is audible)

"I really want this to come out right for the team. My task is due next week. I don't want to let you guys down. I know Joe can do the customer needs analysis. But right now, we're going nowhere, and if he vents on me again I'm gonna quit the team."

Nestor

"Lisa, I hear what you are saying. But what do you want to do about it?"

Step Two External: Point Out Discrepancies. Nestor is confronting Lisa with the fact that she keeps talking about the issue but not finding an alternative.

Lisa first responded to this by simply restating the problem,

"Well, Joe is just not doing his work so I can't do mine."

Nestor,

"Okay that defines the problem. Now how do we find a solution?"

Lisa

"Well, I don't want to have to report to the professor to get Joe to do more of the work with me."

Nestor,

"Yeah, I get that. That is what you don't want as a solution. But what would work better for you than that?"

Step Two Repeated: Pointing out External Discrepancies

Nestor is confronting Lisa with the discrepancy that she wants the problem solved, but she doesn't want to bring in the professor to offer his help or guidance.

Lisa

"Well, I don't know. Maybe it would be better to deal with Joe in a team meeting. It's really a team issue as well as a problem between me and Joe isn't it? After all, if I don't get my work done because of Joe, it won't look good, and the whole team fails."

Nestor,

"So how do you feel about bringing this up in a team meeting? If Joe gave you heat when you brought it up before, do you think you will have it any easier within a group when you bring this up?"

Lisa, leaning back in her seat and slumping her shoulders

"Oh wow, I didn't think of that . . . You know, maybe I just don't fit this team and I'd better get reassigned before the project gets too far on."

Nestor,

"Yeah, I'd want out too if it was me. I can see he really blew it and you don't want to have him vent like that again. How about if it's an agenda item and I bring it up as coordinator. I think Sunil, will get with us on this in a way that doesn't embarrass Joe. She's a great team leader. I've seen her handle this kind of stuff before and not crash."

Step Three: Work toward rebalancing discrepancies. This closes the gap of the discrepancy between Lisa's need for a solution and her need to not involve the professor at this point.

Lisa,

"Yup, she did a great job last year with that team that almost fell apart. I can work with that."

Nestor,

"Okay, if you put the item on the agenda as a general discussion. I'll follow up with Sunil and let her know we might have task completion fall out if we don't cover the agenda item."

Lisa,

"I'll go enter the agenda item right after I leave here, I don't want to take a chance I'll get busy and forget."

Nestor,

"Okay, I'm putting it in my phone now to tag Sunil when I see her in lab tomorrow."

Lisa,

"Thanks, Nestor. This is a little over and above your task coordinator job and I really appreciate it."

Nestor,

"No problem, I've been there and I get it."

Step Four: Evaluate Outcomes

Lisa and Nestor can evaluate outcomes as they place the item for discussion on the agenda and Nestor talks to Sunil. If these outcomes are completed they don't need further evaluation. If the agenda for the meeting is already full or Sunil has a different opinion about how to deal with the issue then Nestor and Lisa will re-evaluate outcomes.

This dialog, though brief, illustrated both confrontation that was not productive and confrontation that was productive.

The confrontation between Joe and Lisa was not productive.

- Lisa identified her perception of the discrepancy that Joe was not doing the work he had agreed to do and that she was then unable to do the work she had agreed to do.

- Lisa pointed out this discrepancy to Joe. But she was unable to point this out as a discrepancy in shared teamwork agreements and task completion thus far. She did what most of us would do. She pointed this out as a discrepancy between Joe's stated task agreement and Joe's actual task completion. Putting Joe on the defensive.
- Joe was not receptive to the information as Lisa transmitted it to him. The information got lost in the interpersonal weather between two team members.
- So there was no agreement reached as to how the discrepancy could be accurately defined or resolved with agreed upon actions.

The confrontation between Nestor and Lisa was productive.

- Nestor was able to identify the discrepancy between Lisa's need to get Joe to do more work and her need to keep the supervising professor out of the situation.
- Lisa was receptive to Nestor pointing out the discrepancy because he used nonverbal and verbal communication skills that were direct, clear, and affectively balanced. So they were both able to choose actions to balance the discrepancy rather than co creating more interpersonal or technical content discrepancies.
- Lisa acted on her agreement by adding a work timeline and individual follow on discussion to the agenda.
- They are each planning on acting on their agreement by ensuring the discussion occurs early in the meeting.

The discussion that ensues over this agenda item in the team meeting is a useful example of both unproductive and productive Conflict Negotiation.

HOW TO CLEAR OBSTACLES USING CONFLICT NEGOTIATION

Much has been written and practiced about conflict negotiation over the course of human history. One could almost say that human history is defined by conflict negotiation and that societies actually rise and fall based on their abilities to manage conflict within their own borders and across the boundaries of other societies. As an engineer you do not have to be an expert in this area to deal with conflict successfully in technical settings.

But you can minimize the time you spend handling conflict by understanding a few key aspects of conflict and conflict negotiation.

The *first key concept* is that conflict, like other expressions of communication, happens cognitively and also in that realm of the 87% nonverbal, affect based communication we have been hammering home time-and-time again through out this book. In other words, conflict is a physical, sensory process even when you are not using a trebuchet to bring down somebody's castle.

Fortunately the nonverbal aspects of conflict negotiation are a well-researched area of human communication. Most notably by Dr. John Gottman, who studied the physiology of conflict in laboratory settings.

He found measurable body changes that occurred consistently and across cultural groups, in heart rate, saliva, muscle acidity, muscle tone, breath rate, cognition and perceptual clarity and associative and perceptual memory abilities during conflict, and conflict negotiation. These changes occurred even during medium stress conflict negotiation between couples on both pragmatic and interpersonal issues (Gottman et al., 2001).

The *second key concept* for you to understand about negotiating conflict in *engineering contexts*—is that you are *in a relationship* with the engineers you work with, individually and in teams. These are work relationships, In the same vein as Gottman, and very powerful relationships at that. These relationships are powerful because your shared performance with other engineers impacts promotions, increased salary levels, and job retention. These relationships are powerful because if you amplify each others' errors rather than finding ways to work clearly together, then your designs and operations can be compromised at best and endanger public safety at worst.

This means the relationships between you and your teammates or peers have tremendous power in your own lives and in the lives of members of the societies your profession serves. That is a basic definition of relationship—a consistent daily connection based on mutual goals and shared, life-defining resources. *So when you have professional conflicts in relationships that occur in engineering contexts they can induce emotions that are every bit as disorienting as those you experience at home with family and friends.*

Putting these two key concepts together, we refer back to Dr. Gottman and peers who found that affective flooding happens even in low to medium stress stages of an argument between two people, especially when those two people are in a relationship involving important mutual resources and shared goals.

In other words, even in low to medium impact conflict moments—the minute someone feels their own perspectives, goals or outcomes threatened—their cognitive processes become flooded with associative and sensory affective stimuli. Their emotion begins to challenge clear processing capacities that support constructive cognition and behavior.

This challenge results in

Either

The brain responds to affective stimulus with a focus on one-point outcomes instead of subtle, systemic changes and the perspectives of others.

Simultaneously, heightened stimulus response is funneled into movement tracking, selfdefense and fight-or-flight responses—and not into cognitive or verbal clarity.

Or

Heightened stimulus response is responded to by the brain with an intensification of sensory affective cue reception, creating an over attentiveness to the statements and reactions of others and an inability to stay anchored in one's own experience. This response also mutes left-brain cognitive focus, decisional language choices and behavioral action capabilities.

Wow! Did you get that or did it pass by too quickly? What we are saying is that early in a conflict, the minute you sense that your position or needs are threatened—whether those needs are technical or interpersonal—you move into a brain processing pattern that minimizes clear, technical cognition and maximizes high volume affect reception.

Now apply the two key concepts of conflict negotiation we mentioned at the beginning of this chapter to the basis of conflict dynamics we just clarified for you.

First, understand that the core of conflict happens when discrepancies between differing perspectives, needs, and resources block or interrupt engineering tasks. This broad and generalized statement covers every kind of conflict from deciding what software to use for an analysis, to an employee being fired, or a project being dissolved. The following quote from a respected conflict negotiator gives you a good idea of the complexity of conflict negotiation in any such setting.

> From the inside, when people become so focused on resolving the one issue most important to them, they pay less attention to what is going on around them, and may not have the ability to consider the perspectives of others. From the outside, the types of issues that may create conflicts for people may have a much broader social origin in structural inequalities and knowing these "variables" can help us be better conflict workers. In other words, despite appearance, people are highly motivated and have the tools, but most need some level of assistance in gaining the necessary perspectives and authority to resolve conflicts.
>
> The breadth of research and theory on conflict points to the combination of physiological, cognitive, emotional, social, and contextual sources and implications.
>
> —Hayes (2011)

Second, understand that the conflict negotiation structure we offer is based on this wisdom and adapted for use in engineering settings.

Lisa and Joe could have used this structure during their conflict in the lab. They needed this structure because they were both unwilling and/or unable to identify technical and interpersonal, as well as internal and external, discrepancies. Lisa was disconnected from her feelings—assuming she was only communicating technical information to Joe, unaware of how her nonverbal signals were being transmitted around the technical content. Joe was flooded by overdriven affect. His cognition, language skills, and behavioral choices abandoned technical aspects of the exchange. His communication became purely interpersonal and was expressed as selfdefense rather than finding a balance of technical and interpersonal communication. This was unproductive conflict because it disconnected and overdrove technical priorities. Joe and Lisa's discrepancies polarized and then entrenched, creating roadblocks to engineering communication and technical performance. Productive conflict negotiation skills could have prevented those outcomes.

Unproductive conflict means that affective, cognitive, and behavioral processes have run off the rails. So it becomes a very good choice to use the conflict negotiation structure presented in the box below to get back to constructive conflict quickly. In our presentation of this structure we give you rules to follow rather than steps to take, because respecting any one of these rules enables you to find stability within a conflict negotiation and quickly move it back into the realm of the constructive. Each rule can be applied at any point in the conflict to shift the dynamic of the conflict.

CONFLICT NEGOTIATION RULES

Rule One

When you are in conflict, stabilize and then keep re-stabilizing yourself. Find where in your body you are feeling distressed. Tag the breath, posture, voice tone, heart rate, and body language cues you are expressing that show you that you are upset.

- Practice RECAP toward yourself so you *re-anchor in your own body.*
- Note for yourself what you would like the other person or the group to stop or start doing that would *ease your own conflict response.*

Rule Two

Observe the other person or group members carefully so that you do not escalate their interpersonal responses by your own nonverbal and verbal choices.

- Tag their *nonverbal* responses.
- Identify if they are afraid, angry, hurt, taking charge, disconnecting, or overdriving.

- Listen carefully to technical and interpersonal content, mark discrepancies and gaps in needs and available resources.

Rule Three

Back off and give yourself a break from the action.

- Ask for a Time Out when things get too heated.
- Make a change in your physical position—move your body, leave the room, get a drink of water—this can interrupt escalating thoughts and feelings that can disconnect or overdrive communication during conflict.

Rule Four

Focus on the positives and strengths of each of your teammates or your individual peers. Focus on your own positives and strengths. Use the positives about your teammates as you *highlight situational discrepancies to them while using constructive verbal communication choices.*

- Recall a time when you did reach a constructive solution out of conflict with these or other peers.
- Highlight strengths and positive contributions of your peers.
- Trust your own perspective and your teammates emotional intelligence enough to highlight discrepancies with productive verbal and nonverbal feedback.
- Reflect on your own and others polarized and entrenched discrepancies with an understanding that every negative position hides a positive position. The energy you are using to hold polarized positions and maintain entrenched patterns can be released and recycled into momentum for constructive outcomes.

Rule Five

Do not criticize, do not use sarcasm, do not use nonverbal body postures, voice tone and facial expressions that broadcast anger, hurt, disgust, contempt, blame, or disconnect.

These emotions, when used during communication in conflict, have been proven by the extensive research of Dr. John Gottman and peers to be 94% accurate in predicting very low relationship satisfaction and very high rates of relationship dissolution.

This is important because teams are relationships that involve a consistent daily connection based on mutual goals and shared, life-defining resources. So respect the relationship by

- Using "I" statements that are noncritical of others to state your view of the problem.
- Suggesting practical steps toward problem discussion and resolution rather than redefining the problem over and over again.
- Being ready to apologize and try again when you can see you have alienated a teammate—*even if you do not feel your nonverbals, words or behaviors were intended to be out of balance or hurtful.*
- Stating your own needs in constructive terms and advocating for solutions that are sustainable for both you, the team, or the group.

Engineering Project Conflict Negotiation Scenario II

The best way to understand the use of this structure is to see it unfold in annotated detail.

Here is a team meeting where the whole team, Lisa, Nestor, Sunil, Chun and Joe, try to negotiate conflict around their engineering task roadblock.

Lisa, Nestor, Sunil, Joe, and Chun are a five-person project team that is developing a General Autonomous Robotic Device with Expert Networked EffectoRs (GARDENER) to be used to tend plants in a home garden. This is a capstone project and done in collaboration with a Community Garden for public schools in their city. It will be coordinated with an educational initiative being done in the schools.

The five engineers on this team are excellent students and have high grade point averages for the courses they have completed in their program. They competed hard to win a place on this team because it exemplifies a prestigious, high profile program their school is known for—one that helps the reputation of the graduates as highly innovative and progressive engineers. In years past the projects have been highly respected examples of both technical innovation and community service. The teams make their final presentations to a group of community leaders, venture capitalists, business developers, and distinguished alumni. This external set of evaluators determines the final scores for the project, and the subsequent award as the year's Most Innovative Product.

In the initial team formation stages they decided to organize along the design characteristics of the system: (Sunil) energy storage, power production, and drive train; (Nestor) communications network; (Joe) navigation and central processing; (Lisa) sensors and effectors; (Chun) chassis and integration. In addition, team members have roles associated with the conceptualization, design, implementation, and operation: (Joe) customer needs analysis and concept of operation definition; (Lisa) system requirements definition and analysis; (Nestor) life cycle cost estimation and analysis; (Sunil) alternative design trade off and selection; (Chun) test and evaluation. To create a balanced team structure, the team designated roles as leader (Sunil), monitor (Chun), coordinator (Nestor), recorder (Lisa) and checker (Joe) for each meeting. These roles are rotated once a month, to divide up the responsibility for assignments over the projected five-month life of the project. Here's the scenario as it unfolds in the next team meeting.

(Conflict Negotiation rules are annotated in italics after each individual team member statement if that statement broke or followed a conflict rule.)

Sunil,

"Okay, we've taken care of Chun's report on chassis and integration possibilities and projections for the development of our test and evaluation plan. He's going to interface with the neighborhood co-op folks to see if their grant funds would cover costs for his basic options so far, then report back to Joe to use as bounds for the costing. Our next item on the agenda is coordinating task performance."

Joe, (leaning back in his chair and crossing his arms, glancing at Lisa)

"That's not a legitimate agenda item for this stage. We covered that in early team meetings. We don't need to go over it again and whoever put that on there is holding up the team."

Breaking Rule Five: Criticizing a teammate, broadcasting nonverbal communication signals that are closed and confrontational.

Sunil, (taking a breath, relaxing her shoulders and including the whole group in her gaze)

"I hear you, Joe. This is going over old territory. Does everybody feel what we agreed on is working? How about we take five minutes to at least check in on this? This *is* something that often needs review mid-project."

Following Rule Five: Using I statements, making constructive suggestions, using nonverbal communication that is stable and open.

Lisa, (leaning forward in her chair, putting her hands on the table, glancing at Joe, raising her voice a notch)

"That's a great idea and if we don't do it then we're holding the team back."

Breaking Rule Five: Using generalized statements, broadcasting nonverbal communication that is assertive and confrontational.

Chun (speaking quietly and calmly, head down, sifting through papers on the desk)

"I'm fine with checking in. There's a form in our rubric here for team check-ins about six weeks into the project. When we couldn't agree on who would be first team leader in the rotation, we followed the instructions on the rubric and it really helped. So let's follow what they gave us for this item."

Following Rule Four: Remembering a time when conflict was productive.

Joe, (putting his arms down, shoulders slumped, looking at Chun)

"Alright, if it's in the rubric I guess we have to do it."

Nestor, (looking at Joe with a nod and a smile, opening his posture and relaxing back in his chair)

"Thanks for taking the ball on that one, Joe. I agree with you, following the rubric is a requirement not an option."

Sunil, Chun and Nestor laugh. Lisa smiles a wobbly smile, Joe relaxes a bit.

Sunil,

"Nestor, why don't you facilitate the check in?"

Nestor, (sitting up straight, including the whole team in his gaze)

"Okay, I'll take it on. But let's follow the conflict structure if things get rough, all right? Addison over on Team B said they used it last week and now they seem like they are moving way ahead of us. It really upped their productivity with each other."

Following Rule Five: Using 'I' statements, making constructive suggestions.

Chun sifts through the rubric to find original project agreements.

Lisa, (tapping her toe loudly under the table, hunching over her laptop and beginning to open work files, glances in Joe's direction and sighs heavily)

"We wouldn't need the structure if everyone would just do their job. The next iteration of the customer needs analysis is central and I don't see it yet. I've got all my files right here and prepped and ready to go but I can't finalize the systems requirement statements without the customer analysis feedback happening first. Everybody knows that."

Breaking Rule Five: Using criticism, sarcasm, disgust, and blame. Broadcasting nonverbal communication that is assertive and confrontational.

Dead silence ensues.

We now interrupt this dialog to show you . . .

This is your Brain. This is your Brain on Conflict.

Joe is experiencing the first unconstructive reaction to the heightened conflict stimulus that we mentioned earlier in this chapter.

Lisa is experiencing the second unconstructive reaction to heightened conflict stimulus response that we mentioned earlier in this chapter.

Having either response is not really a problem. It is *what you do with the response that can become a problem*, especially when you are sure, so sure that you are actually focusing solely on technical content and therefore you are not being emotional. Or especially when you are sure, so sure, that a colleague or teammate is acting so outrageously that it is time to slash-and-burn and to heck with technical needs.

Dialog Continued . . .

Sunil glances quickly at Nestor. Joe roles his eyes and goes beet red and folds his arms across his chest. Chun looks down at his papers and shuffles them nervously.

Joe, (sitting up sharply and quickly in his chair, looking directly at Sunil and Nestor)

"Since you two seem to be running this game, can we please talk about anything, *anything* but this right now? How about our choices for network communication and processing? Why don't we talk about that? *That's* the technical core of this project, guys. *That's* where my Smartphone based idea gets us a good grade here, people, and that's what impresses the community and serves the project above all else. Not this silliness about who does what at exactly, the very moment, the absolute on time, turn-on-a-dime minute that somebody else thinks they need something."

Breaking Rules One, Two, and Five—Joe is in affect processing overdrive. He is not internally connected to his own nonverbal or verbal choices. He is not selfstabilizing and he is not observing others. He is using sarcasm, blaming, and criticism. This conflict is now officially unproductive.

More dead silence.

Lisa, (pushing aside her computer, and leaning forward so far she is practically on the table, face diffused with frustration)

"Are you kidding? Are you seriously, really kidding? The choice of network communication and central processing is the core of the project? There *is* no project without meeting customer need. Yup, you heard me. No. Project. Without. Meeting. Customer. Needs."

Breaking Rules One to Five. Jumping into the deep end of the pool with Joe to swim in disconnected and overdriven affect.

Chun pushes a piece of paper across the table to Nestor, he points at a section of the paper. Nestor reads it quickly and clears his throat nervously.

Nestor, (leaning back in his chair, taking a breath, relaxing his shoulders, placing his feet on the floor)

"TIME OUT. I am going to interrupt everybody and I am asking for seven minutes, by the clock, when nobody talks. Silence. Please. We can get through this.

Joe, (popping out of his chair and heading for the door)

"I hear you, Nestor. I'm following the Time Out rule that says step outside. I'll be back."

Following Rule Three: Taking a time out and stepping outside to diffuse conflict

Following Rule Five: Making constructive "I" statements.

When you are in Conflict—Step Outside

Research on brain response to stress arousal shows that it is possible to use nonverbal, sensory cues to selfsoothe. You can use these cues and get yourself back on track any time you are acting in ways that damage your relationships during conflict (Greenspan, 2001; White, 2008). Additional research has found that an almost instant way to practice this selfsoothing is to step outside (Cohen, 2007; Murchie, 1977).

So when you are in conflict, step outside.

Find a few potted plants in an atrium, step out on a balcony, walk into a courtyard, or step out the front door. Take a breath, connect with the natural environment and allow your body to be soothed.

Why?

Your brain is wired with more than forty-eight additional senses that support and inform the standard five senses that drive your emotions and reasoning capabilities. These additional senses are wired to feel good outside. They were originally used to read the landscape for cues that led to food, shelter, warmth and community and thus predate neo cortex development of cognitive function in language and reason. So when you step outside and use these senses, you have direct input to affective and perceptual brain processes that strengthen selfsoothing. This slows down the emotional and cognitive loops that go haywire in the human brain during verbal conflict, clearing the path for balanced reason (Cohen, 2007).

When you Step Outside

Your body feels and breathes in fresh air and this relaxes your over stimulated affective responses very quickly.

You see leaves, rocks, grass and dirt, and hear wind and birds instead of people, walls, computers and lights. These more natural sensory stimuli balance limbic brain response with executive brain response. It becomes easier to balance feeling and thinking.

When Joe walked down the stairs, stepped outside and walked around the courtyard. His breath rate slowed, his eyes and brain engaged with the landscape rather than his feelings and angry thoughts. He was able to re-establish cognitive and affective balance and thus technical and interpersonal balance.

While Joe was outside . . .

Lisa, (opening her backpack and taking out a cough drop and a flat, palm sized stone)

"You guys, I am so sorry. I'm just gonna sit here a minute and work the worry stone until I get my head back. (Sits back in her chair and places the stone in her palm, pops the cough drop in her mouth and gazes out the window. She takes deep breaths.)

Following Rule One: Practicing RECAP toward herself and selfstabilizing.

Following Rule Five: Apologizing when necessary.

Nestor watches the clock. Sunil picks up her phone and plays some Sudoku. Chun takes out a harmonica and starts to play softly. That makes Sunil laugh and Nestor teases Chun about putting him on YouTube. Lisa looks at everybody and really smiles for the first time. Joe comes back in the room.

He nods curtly at Lisa. He sits down and shoves his chair back away from the table.

Joe

"Okay, Sunil. I'll do the check in. But I wanna put something more on the agenda."

Following Rule Five: "I" statements that move progress forward and constructive suggestions.

Sunil nods her head encouragingly and hands Joe the pad with agenda items,

"Go ahead Joe, your input gave us great momentum when we were forming. Just keep it short and keep it focused."

Following Rule Four: Highlighting strengths and positives of a team member.

Following Rule Five: Making constructive suggestions.

Joe (visibly making an effort not to glance at Lisa, not to cross his arms or raise his voice)

"I've got a resource for navigation and control that could make this whole thing really work. My buddy over at the robotics lab was giving me feedback on my concepts and he showed me an interface between sensors and processors that is the most user friendly I've seen so far. Since there will be teachers and students using this thing I think we need user-friendly and we haven't dealt with that at all yet. So I want this on the agenda and I want the check-in not to run over time."

Following Rule Four: Trust your own perspective and your teammates emotional intelligence enough to highlight discrepancies with productive verbal and nonverbal feedback.

Following Rule Five: Making constructive suggestions.

Alert—Roadblock Becoming Constructive Momentum—Joe stabilized his own affect enough to be able to forward technical design stages and his

own contribution to the project. He rebalanced technical and inter-personal aspects of his engineering communication with the team.

Lisa, (crossing her arms over her chest, then pressing her worry stone, coughing, then sitting back and opening her posture, smiling at Joe)

"That takes a lot of time, Joe. I am glad you're adding that to the project but I'm getting behind. I do need your analysis. How did you have the time to do all that but not the next iteration of needs analysis?"

Following Rule Four: Trust your own perspective and your teammates emotional intelligence enough to highlight discrepancies with productive verbal and nonverbal feedback.

Alert—Roadblock Becoming Constructive Momentum—Lisa stabilized her own affect, acknowledged Joe's technical contribution and construc-tively asked for team cooperation. She rebalanced technical and inter-personal aspects of her engineering communication with Joe and the team.

Joe, (crossing arms back over his chest, using a curt voice tone)

"That's not the point, the point is this product needs . . . "

Sunil, (sitting up straight in her chair, speaking in a firm voice)

"I'm going to interrupt here. Let's stay on task. Joe, great work. Add your item to the agenda. It's a good thing to discuss. Now, let's do the check in."

Nestor, (leaning forward slightly in his chair, smiling at Joe, nodding encouragingly)

"Okay, Joe. I just want to ask, buddy. That controls work takes hours of time. I don't get it. How come you have time for that but not the customer needs analysis? I'm waiting for Lisa's system requirements and she can't do those until you do your work."

Following Rule Four: Highlighting discrepancies with constructive nonverbal and verbal choices.

Joe, (sitting back, taking a breath, clearing his throat, going red in the face)

"Sorry guys, I didn't get the customer analysis done as easily as I thought I would. My classes run from nine to three on Monday, Wednesday, and Friday. My student council position takes up most of Tuesday and Thursday and I work most evenings. All I have for time to make phone calls is lunch. All the schools we are working with close at three. And those teachers over there, man. They don't ever come to the point. You get them on the phone and hear how great the kids are and how happy they are to work with us and all you want to do is ask a simple question about is does what we have defined meet what they need at their end. Then Lisa's been calling me and texting me and emailing me so any time I have at noon time I am trying to also answer her back. But I can't come up with a voice mail or text that doesn't show how hacked off at her I am so I send nothing. I should have asked for some feedback sooner. Sorry guys, sorry Lisa."

Following all the conflict negotiation rules—Completion of roadblock transformation into team momentum.

Lisa, (sitting forward in her chair, looking at Joe with a mystified expression)

"You did all that and the controls basics? Why didn't you say something?"

Following Rule Four: Highlighting discrepancies with constructive verbal and nonverbal choices.

Chun, (looking at everyone and shrugging his shoulders apologetically, then smiling)

"Let's not go there again or I'll get a rash, Lisa. Joe, I have an Aunt who works in one of the schools. I'll get with her and see if there is somebody who can be a point person for us and contact Joe at a time when he is not at work, school or designing a new interplanetary navigational control system."

Following Rule Five: "I" statements that move progress forward and constructive suggestions.

Sunil

"Okay guys, now that I'm drooling relief slime over here. Does anybody else have issues with task coordination? Any other bugs we need to work out? I suggest we put a check-in at the beginning of each meeting. That way things don't get so gnarly next time. Joe, you can be the time limit keeper on that and Lisa, you can coordinate it. How's that sound?"

Nestor, Chun, Lisa, and Joe

"Yeah. Sounds fine. Let's move on. I'm good to go with that."

Meeting Adjourned

Try This

Go through this dialog and find one example of each microskill presented in the chapters in this book.

Go through this dialog and highlight nonverbal behaviors that demonstrate affect disconnect, overdrive, or affect that is balanced and supports both interpersonal and technical communication clarity.

Go through this dialog and think about how you would think, feel and act if you were Lisa, Nestor, Joe, Sunil, and Chun.

Think of a recent minor or more difficult confrontation or conflict you experienced in the practice of engineering. Identify one conflict rule you followed or broke.

The confrontations and conflict negotiations that happened in this dialog transformed unproductive discrepancies into productive creative resources. They were then used to drive intentional technical and interpersonal communication exchanges toward constructive outcomes.

The five members of this team practiced effective engineering communication skills in an authentic engineering setting and experienced a positive impact on their communications and engineering outcomes. When you can do this yourself, you also will be approaching a practical utility in both your technical skills and the interpersonal fields in which they are transmitted and received.

CHAPTER 16

BECOMING AN INTENTIONAL ENGINEER

Intentionality means that you can understand and practice the microskills. You can use individual microskill tabs in our Communication Microskills Model in unique and diverse pairings across strands of changing engineering situations and settings. You have demonstrated that you can drive your own responses and the responses of others even in stressful or creative engineering communication situations.

When you become an intentional engineering communicator, you have demonstrated that you can excel in a proficiency that your fellow engineers think is a priority in your profession. We now offer you a step above excellence, adding additional value to your engineering proficiency. We offer you the capacity of intentional engineering.

Intentional engineering means working with others to get results, being able to influence others to create a shared goal, and reaching that goal. It means taking responsibility to get results not only from yourself, but also from others. It means becoming a proficient and competent engineer who is valued within the shared system of primary social needs and services that are at the foundation of your profession. It means you know how to behave as a proactively contributing member of that social system and its unique communication needs, using both technical and interpersonal modes of communication effectively for the needs of that system.

Effective Interpersonal and Team Communication Skills for Engineers, by Clifford A. Whitcomb and Leslie E. Whitcomb.
© 2013 by The Institute of Electrical and Electronics Engineers, Inc. Published by 2013 John Wiley & Sons, Inc.

This intentionality is expressed as an ability to constructively shape the flow of technical and interpersonal communications during your teamwork and project operations. You are capable of using interpersonal dynamics to enrich technical processes, clarifying them and creating potential for their most accurate expression. *You are able to anticipate and modulate conceiving, designing, implementing, and operating challenges based on the interpersonal and technical patterns of communication you see ongoing in task processing in a variety of engineering situations and settings.* By any frame of reference, this is engineering excellence.

Bringing this value to your profession is important for two reasons. The first is the fact that throughout your career you must be able to let others know of the technical aspects related to your engineering efforts in situations that will not always be primarily technical on the Space, Face, and Place spectrum. The second is the reality that present-day engineering happens largely in teams and that local and regional teams are impacted by global forces. Being a successful member of a team and providing proactive team leadership in this context necessitates that you are capable of communicating both interpersonally and technically in fluid and adaptable ways. As you learn to do this, you will be taking intentional action in your profession. You will become an intentional engineer and be capable of consistently practicing intentional engineering.

You become an intentional engineer, capable of intentional engineering, by practicing the modules in this book in engineering contexts with enough fidelity that you become fluent in their usage.

You accept that communication skills shape engineering outcomes. You internalize the fact that outcomes are dependent on communication skills because engineering happens in shared technical and nontechnical interpersonal exchanges. You know the difference between talking *about* communication and actually being an effective communicator. You know how to practice *doable* proficiencies in this foundational area of successful engineering. You not only know how to think about communication, you are an effective communicator.

You are an effective communicator because you know how to identify the shared interpersonal field through which all technical and nontechnical communications are personally transmitted and received. You can use your skills in communication exchanges to effectively get your message through to another person, or to a team, because you can read the surrounding interpersonal weather accurately and respond to that atmosphere with relevance and balance.

You practice self-understanding in your own natural style of communication and with your own form of emotional intelligence. You know how to modulate your own and others' affect and are able to do so even during stressful conflict negotiations. If the outcome is not as predicted or expected, you are able to move to alternate communication and technical skills and strategies that get the results you need. You know how to be intentional and successful while you learn to lever interpersonal dynamics in support of your technical choices as you conceive, design, implement, operate products and systems and share your work in interface with society.

You have learned proficiency in each microskill tab on the Communication Microskills Model and you can use pairs and recombinations of microskill tabs easily in a broad diversity of engineering situations and settings. You are capable of exchanging your own DNA strands of microskill tab competence with that of others in even stressful and creative engineering project tasks and cycles.

Your competence with the model enables you to leverage interpersonal dynamics from being potential obstacles blocking excellence in engineering into open flows of momentum toward excellence in your profession. You can architect this excellence because you can identify and transform obstructive communication exchange patterns of yourself and others. This mastery means you have intentional competency not only in engineering communication but also in engineering as a social and professional dynamic that shapes the course of human endeavor.

This robust intentional competency allows you to ensure that a team communicates in specific ways to resolve a conflict, converge on a design solution, balance the effort of individual, group and wider systems efforts and maintain communication exchanges that keep a design process moving forward. This intentional competency means you are able to protect your own interests and input on a project and work smoothly with others to ensure your contributions are received and accepted. It means your own and your team's engineering proficiencies have a shot at making an impact in a complex global engineering context. Not only are you driving your engineering communications instead of being driven by them, but you are also shaping the road you travel as an engineering professional and contributing a constructive, creative path for others to follow.

At this level of proficiency, you have mastered engineering communication skills to the point that you will be able to teach team members, peers, and colleagues how they, too, can become better engineers through the application of communication skills in their technical and nontechnical engineering cycles. Not only will you be an intentional engineer capable of practicing intentional engineering. You will be able to model intentionality in engineering for others and drive the innovative edge of this crucial proficiency as a service to your profession and to society. Congratulations on your achievement! This is not only an engineering achievement. This is a human achievement of no small standing. With this achievement, you will be able to not only improve your skills as an engineer but also improve society as a communicating and functioning system of shared social resources used for the common good.

BIBLIOGRAPHY

ABET. "Curriculum Criteria for Accreditation". Internet: http://abet.org, August 3, 2011.

Birdwhistell, R. (1970). *Kinesics and Context*. University of Pennsylvania Press, Philadelphia.

Bregman, P. (2012). The Emotional Adventure of Leadership, *Harvard Business Review Blog Network*, June 14. Available at: http://blogs.hbr.org/bregman/2012/06/the-emotional-adventure-of-lea.html.

CDIO. "Mission Statement". Internet: http://cdio.org, August 3, 2011.

Crawley, E. F., Malmqvist, J., Östlund, S., and Brodeur, D. R. (2007). *Rethinking Engineering Education: The CDIO Approach*, Springer-Verlag, New York.

Crawley, E.T., Malmqvist, J., Lucas, W.A., and Brodeur, D.R. (2011). The CDIO Syllabus v2.0, An Updated Statement of Goals for Engineering Education. Proceedings of the 7th International CDIO Conference, Technical University of Denmark, Copenhagen, June 20–23.

Druskat, V. U. and Wolff, S. B. (2008). Building the emotional intelligence of groups. *Harvard Business Review*, 79(3), 81–90.

Egan, G. (2009). *The Skilled Helper*, 9th edition. Brooks/Cole, Cengage Learning. Belmont, CA. http://books.google.com/books?hl=en&lr=&id=yfycFi3iT4oC&oi=fnd&pg=PR3&dq=egan+skilled+helper&ots=xRNEg5AuZ_&sig=I1hOcQwpWsnvTC_tlE1xog3ue3Q

Effective Interpersonal and Team Communication Skills for Engineers, by Clifford A. Whitcomb and Leslie E. Whitcomb.
© 2013 by The Institute of Electrical and Electronics Engineers, Inc. Published by 2013 John Wiley & Sons, Inc.

139

Ekman, P. (2007). *Emotions revealed: Recognizing faces and feelings to improve communication and emotional life (2nd ed.)*. New York: Times Books.

Gottman, J., Levenson, R., and Woodin, E. (2001). Facial expression in marital conflict. *Journal of Family Communication*, **I** (1), 37–57.

Greenspan, S. I., De Gangi, G., and Weider, S. (2001). The Functional Emotional Assessment Scale (FEAS); For Infancy and Early Childhood. Bethesda, MD: Interdisciplinary Council on Developmental and Learning Disorders.

Hayes, S. (2011). Understanding conflict resolution from the inside out or why 800 pound gorillas aren't great mediators. New Opportunities for Peacemaking 2011 Dispute Resolution Section Annual Meeting and Conference (pp. II1–II8). Cary, NC: North Carolina Bar Association Foundation.

Knapp, M. and Hall, J. (2009). *Nonverbal Communication in Human Interaction*. Kentucky: Wadsworth Publishing.

Lasley-Hunter, B. and Preston, A. (2011). Systems Planning, Research, Development and Engineering (SPRDE) Workforce Competency Assessment Report. Center for Naval Analyses. Alexandria, VA. June 2011. Work was created in the performance of Federal Government Contract Number N00014-05-D-0500. Distribution limited to DOD agencies.

McDonald, A. and Hansen, J. (2009). "Truth, Lies, and O-Rings: Inside the Space Shuttle Challenger Disaster", Gainesville; University Press of Florida.

Mehrabian, A. *Nonverbal Communication*. Aldine Transaction, 2007.

Niewoehner, R. (2011). CDIO Syllabus Survey: Systems Engineering an Engineering Education for Government. Proceedings of the 7th International CDIO Conference, Technical University of Denmark, Copenhagen, June 20–23.

Norman, D.A. (2004). *Emotional Design*. New York: Basic Books.

Ratey, J. (2002). *A User's Guide to the Brain: Perception, Attention, and the Four Theaters of the Brain*. London: Vintage Press.

Tjan, A. (2012). How Leaders Become Self-Aware, *Harvard Business Review Blog Network*, July 19. Available at: http://blogs.hbr.org/tjan/2012/07/how-leaders-become-self-aware.html.

Tronick, E. (2007). *The Neurobehavioral and Social-Emotional Development of Infants and Children*. New York, N.Y.: Norton.

INDEX

Effective Interpersonal and Team Communication Skills for Engineers, by Clifford A. Whitcomb and
Leslie E. Whitcomb.
© 2013 by The Institute of Electrical and Electronics Engineers, Inc. Published by 2013 John Wiley & Sons, Inc.

Printed in the United States of America
ED-02-19-13